Agricultural Nanotechnology

NIPA® GENX ELECTRONIC RESOURCES & SOLUTIONS P. LTD.
New Delhi-110 034

About the Authors

Nintu Mandal

Nintu Mandal completed his M.Sc. and Ph.D from Division of Soil Science and Agricultural Chemistry, ICAR-Indian Agricultural Research Institute (IARI), New Delhi in the year 2010 and 2015, respectively. During his M.Sc. he was getting ICAR Junior Research Fellowship followed by ICAR Senior Research Fellowship, IARI merit scholarship and UGC Junior Research and DST INSPIRE Fellowship during his Ph.D. tenure. He received Indian Society of Soil Science (ISSS) Zonal Award (North Zone) (2011) for best presentation of M.Sc. Dissertation. He received Indian Society of Soil Science (ISSS) Best Doctoral Thesis Award 2015 for outstanding Doctoral Research. He is recipient of Dr. S.P. Raychaudhary Gold Medal for Outstanding Doctoral Research. He published 10 research/review papers in several reputed national/ international journals. He also co-authored in 03 book chapters. Currently he is working as an Assistant Professor cum Junior Scientist in the Department of Soil Science and Agricultural Chemistry, BAU, Sabour, Bhagalpur, Bihar. His current research areas are Nanoformulations for increasing input use efficiency, Soil Chemistry and Clay Mineralogy.

Abir Dey

Abir Dey completed his M.Sc. and Ph.D. Degree from Division of Soil Science and Agricultural Chemistry, ICAR-Indian Agricultural Research Institute (IARI), New Delhi in the year 2011 and 2017, respectively. During his M.Sc. he was getting ICAR Junior Research Fellowship followed by ICAR Senior Research Fellowship, IARI merit scholarship and UGC Junior Research Fellowship during his Ph.D. tenure. He received Indian Society of Soil Science (ISSS) Zonal Award (North Zone) (2012) for best presentation of M.Sc. Dissertation. He received The Golden Jubilee Award for Outstanding Doctoral Research in Fertilizer Usage by Fertilizer Association of India (2017), The Mosaic Company Foundation Outstanding Doctoral Research Award in Area of Plant Nutrition (2017) and ISSS- Commendation Certificate for Best Presentation of Doctoral Research Work done in Soil Science (2017). He published 12 research/review papers in several reputed national/international journals. He also co-authored in 03 book chapters. Currently he is working as an Agricultural Research Scientist in the Division of Soil Science and Agricultural Chemistry, ICAR-IARI, New Delhi. His current research areas are conservation agriculture, soil carbon dynamics, and quality of soil organic matter, soil fertility and plant nutrition.

Anupam Das

Anupam Das, is working as an Assistant Professor-cum-Junior Scientist in the Department of Soil Science and Agricultural Chemistry in the Bihar Agricultural University, Sabour, Bhagalpur, Bihar, India. He has obtained his B.Sc. (Ag.) Hons. Degree from Bidhan Chandra KrishiViswavidyalaya in 2009 and M.Sc. (Ag.) with Gold Medal in Soil Science and Agricultural Chemistry from Banaras Hindu University in 2011. He is the recipient of Junior Research Fellowship of ICAR during Master's Degree and Senior Research Fellowship by ICAR and INSPIRE Fellowship by Department of Science and Technology, Govt. of India during persuasion of Doctoral Degree. He has published a no. of research articles of national and international repute.

Vinay Kumar

Vinay Kumar presently working as Assistant Professor cum Junior Scientist in the department of Soil Science and Agricultural Chemistry, Bihar Agricultural University, Sabour, Bhagalpur, Bihar, India. He has obtained his B.Sc. (Ag.) Degree from Mahatma Phule KrishiVidyapith (MPKV), Rahuri, Maharastra in the year 2003. He completed M.Sc. (Ag.) in Soil Science and Agricultural Chemistry from Dr. Rajendra Prasad Central Agricultural University, Pusa, Samastipur, Bihar in 2006. He has published a no. of research articles of national and international repute. He received Young Scientist Award in 2016 from Society for Scientific Development in Agriculture and Technology, Meerut, U.P.

Agricultural Nanotechnology Basics and Practicals

Nintu Mandal
Officer Incharge, Nanoscience and Nanotechnology Unit
Department of Soil Science and Agricultural Chemistry
BAU, Sabour, Bhagalpur, Bihar, India

Abir Dey
Agricultural Research Scientist
Division of Soil Science and Agricultural Chemistry
ICAR-IARI, New Delhi, India

Anupam Das
Assistant Professor cum Junior Scientist
Department of Soil Science and Agricultural Chemistry
Bihar Agricultural University
Sabour, Bhagalpur, Bihar, India.

Vinay Kumar
Assistant Professor cum Junior Scientist
Department Soil Science and Agricultural Chemistry
Bihar Agricultural University
Sabour, Bhagalpur, Bihar, India

NIPA® GENX ELECTRONIC RESOURCES & SOLUTIONS P. LTD.
New Delhi-110 034

NIPA® GENX ELECTRONIC RESOURCES & SOLUTIONS P. LTD.

101,103, Vikas Surya Plaza, CU Block
L.S.C. Market, Pitam Pura, New Delhi-110 034
Ph : +91 11 27341616, 27341717, 27341718
E-mail: newindiapublishingagency@gmail.com
www: www.nipabooks.com

For customer assistance, please contact
Phone: + 91-11-27 34 17 17
Fax: + 91-11- 27 34 16 16
E-Mail: feedbacks@nipabooks.com

ISBN:978-81-96075-49-1

Composed and Designed by NIPA®.

Dedicated to

Dr. Samar Chandra Datta

Emeritus Scientist,

Division of Soil Science and Agricultural Chemistry

ICAR-IARI, New Delhi, India

Prof. Saroj Kumar Sanyal
M.Sc. (IARI), Ph.D. (Cambridge)
FNAAS, FISSS, FAScT
Former Vice-Chancellor
Bidhan Chandra Krishi Viswavidyalaya
Adjunct Professor
Indian Agricultural Research Institute, New Delhi &
Bihar Agricultural University, Sabour, Bihar
Former Chairman, Research Advisory Committee
NBSS & LUP (ICAR), Nagpur &
CRIJAF (ICAR), Barrackpore, West Bengal
Chairman, Scientific Advisory Committee
Tocklai Tea Research Association, Jorhat, Assam
Former Visiting Scientist
International Rice Research Institute, The Philippines
President, Section of Agriculture & Forestry Sciences,
92nd Session of the Indian Science Congress (2004-2005)

Residence
M.Sc. (IARI), Ph.D. (Cambridge)
Saptaparni, 58/3, Ballygunge
Mohanpur-741 252, Nadia, West Bengal
Circular Road
Kolkata-700 019
Tel.: 033-24544694
Mob.: 098302 47072
Email:sksanyalnaip@gmail.com

November 3, 2017

Foreword

It is indeed my pleasure to write the Foreword for the book, entitled **Agricultural Nanotechnology: Basics and Practicals,** which is authored by Dr. Nintu Mandal, Officer-Incharge, Nanoscience & Nanotechnology Unit of the Department of Soil Science and Agricultural Chemistry, Bihar Agricultural University, Sabour, Bhagalpur, Bihar, India. Nanotechnology, which deals with the matter at nanoscale (1-100 nm), is a rather recent technology that is based on the use of materials in a fine state of sub-division, thereby offering more efficient interaction at the desired interface. Indeed, nanotechnology-based applications in agriculture cover, among others, nano-fertilizers, nano-pesticides, fisheries and aquaculture, removal of pollutants through surface-mediated processes, and so on. As with many other modern technologies, nanotechnology, and its application in agriculture needs to be weighed carefully against any environmental footprint and the food bio-safety issues.

Given such background, it is obvious that a large number of experimental studies are being undertaken in this field. However, there seems to be a genuine lack of well-organised text and reference books in the field which will serve as a standard text. The present book by Dr. Nintu Mandal is expected to bridge this gap. Indeed the given book includes experiments pertaining to, among others, the chemical as well as microbial routes of synthesis of several nanomaterials of potential use in agriculture, coupled with their multi-faceted uses.

I find the book very interesting, and sincerely believe that it will serve the purpose of the targeted end-users in good measure.

-Sd
(Saroj Kumar Sanyal)

Preface

Nanotechnology is the cutting edge science dealing with fabrication or manipulation at atomic or molecular scale. Nanomaterials by virtue of its increased surface area coupled with enhanced chemical reactivity are useful in increasing agricultural input use efficiency. Basic understanding and hands on expertise in Nanosynthesis are of utmost importance in agricultural science.

The Book entitled **Agricultural Nanotechnology: Basics and Practicals** covers basic and practical aspects of nanotechnological application in agricultural science. History of Nanotechnology, Basic Concepts, Definitions, Type of nanomaterials and its peculiarities have been described. Synthesis of nanomaterial's through physical, chemical and biological modes have been covered. Characterization of nanomaterial's through microscopic, spectroscopic and other techniques have been elucidated in lucid language.

The authors would like to acknowledge Dr. S.K. Sanyal, former Vice-Chancellor, BCKV, Nadia, West Bengal for writing the Foreword . Authors are grateful to Dr. Samar Chandra Datta, Emeritus Scientist, ICAR-IARI, New Delhi, Prof S.S. Mukhopadhyaya, Former Director Electron Microscopy and Nanosceince Unit, PAU and Dr J.C. Tarafdar, Former ICAR- National Fellow for their constructive suggestions and critical inputs during the preparation of the manuscript.

Authors acknowledges the enormous support with blessings from Prof. (Dr.) Ajoy Kumar Singh, Hon'ble Vice-Chancellor, BAU, Sabour, Bihar.

This book will be helpful for UG and PG students of Agricultural Sciences including Horticultural, Veterinary, Animal and Fisheries disciplines. We hope this manual will be useful to the students of undergraduate and post-graduates of agricultural sciences for developing hands on expertise in nanotechnology.

Authors

Contents

Chapter 1

Nanotechnology Basics

History and Recent Developments

Richard P. Feynman in 1959 suggested that it should be possible to build machines small enough to manufacture objects with atomic precision. His talk *"**There is plenty of Room at the Bottom**"* is widely considered to be the foreshadowing of Nanotechnology. He delivered this talk at the annual meeting of the American Physical Society at California Institute of Technology (Caltech), In 1990, a team of IBM Physicists revealed that they could write (arrange) the letter IBM using 35 individual atoms of Xenon.

In 1992 a book entitled Nanosystems: Molecular Machinery, Manufacturing and Computation was published by Eric Drexler where he outlined a way to manufacture extremely high performance machines out of molecular carbon lattice (diamondoid).

Around year 2000, federal funding for Nanotechnology in the United States began with National Nanotechnology Initiative (NNI). ***NNI defined nanotechnology as dealing with materials with sizes between 1 and 100 nanometers exhibiting novel properties.***

The Government of India, in May 2007 launched a mission on Nano sceince and nano technology (Nano Mission) with an allocation of ₹1,000 crores.

What are Nanoparticles?

The definition of the term nanoparticle is the subject of debate, and is continuing to evolve. Within an Earth science context, Banfield and Zhang (2001) suggested that nanoparticles might be defined based on the size at which fundamental properties differ from those of the corresponding bulk material. According to such a definition, the size range that constitutes a nanoparticle will vary for different materials, but current evidence suggests that this size range often is between roughly 1 nm and a few to several tens of nanometers for Earth materials (Hochella *et al.*, 2008). According to EPA (2007), NP are therefore considered substances that are less than 100 nm in size in more than one dimension. Such particles are minerals that are as small as roughly 1 nm and as large as several tens of nanometers in at least one dimension. Limiting size in one, two, or three dimensions results in a nanosheet (eg. Vernadite), a nanorod (eg. Palygorskite),or a nanoparticle (Ferrihydrite), respectively (Fig. 1).

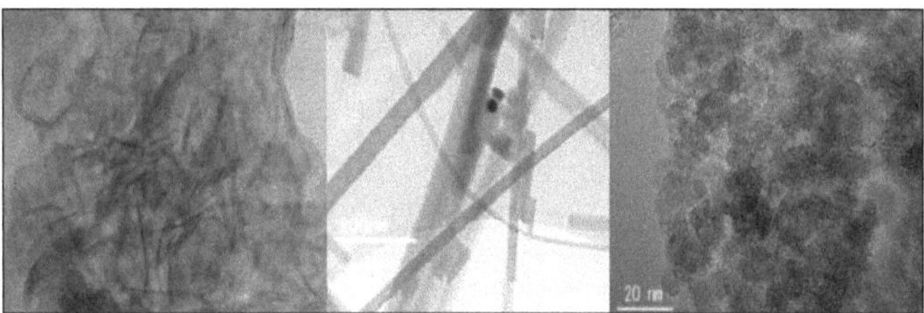

Fig. 1 SEM images of Vernaditen nosheet, Palygorskite nanorods, Ferrihydrite nanoparticles, respectively (Hochella *et al.*, 2008).

Nanoparticles overlap in size with colloids (Fig. 2), which range from 1 nm to 1 mm in diameter. Given that materials are often defined as dissolved by passing through a 0.45 μm (450 nm) filter, nanoparticles are often included in the dissolved fraction, even though they are clearly distinct from molecules and ions. Viruses and many inorganic colloids qualify as nanoparticles. Although bacteria are much too large to be considered as nanoparticles, they may produce nano scale biominerals. It is important to note that nanoparticles are often aggregated into colloidal or larger grains, which can greatly affect their properties with respect to, for example, transport, reactivity, and other geochemical characteristics.

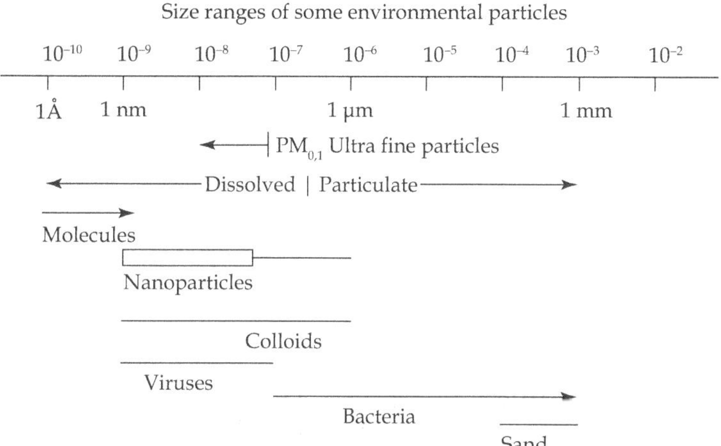

Fig. 2 The size ranges of some environmental particles, including nanoparticle (Maurice, 2009)

Hochella *et al.* (2008) defined nano minerals as minerals such as ferrihydrite that only exist in the nanoparticle size range, or clays that only exist with atleast one dimension in that size range, whereas mineral nanoparticles are minerals that are in the nanosize range, but also exist at larger sizes. Mineral species that can exist as mineral nanoparticles probably include the majority of all known minerals. The fact that nano minerals and mineral nanoparticles may have properties, including stability and reactivity, that change as a function of size makes them fundamentally distinct from larger scale materials.

Classification and Abundance of Nanoparticles

Nanoparticles can be divided into natural and anthropogenic particles (Table 1). The particles can be further separated based on their chemical composition into carbon-containing and inorganic nanoparticles. The C-containing natural nanoparticles are divided into biogenic, geogenic, atmospheric and pyrogenic nanoparticles. Examples of natural nanoparticles are fullerenes and CNT of geogenic or pyrogenic origin, biogenic magnetite or atmospheric aerosols (both organic such as organic acids and inorganic such as sea salt). Anthropogenic nanoparticles can be either inadvertently formed as a by product, mostly during combustion, or produced intentionally due to their particular characteristics. In the latter case, they are often referred to as engineered or manufactured nanoparticles. Examples of engineered nanoparticles are fullerenes and CNT, both pristine and functionalized and metals and metal oxides such as TiO_2 and Ag. Engineered nanoparticles are the main focus of the current research on nanoparticles in the environment, but some of them occur also naturally, e.g. as inorganic oxides or fullerenes (Nowak and Buchelli, 2007).

Table 1 Classification of nanoparticles (Nowak and Buchelli, 2007)

		Formation		**Examples**
Natural	C-containing	Biogenic	Organic colloids	Humic, fulvic acids
			Organisms	Viruses
		Geogenic	Soot	Fullerenes
		Atmospheric	Aerosols	Organic acids
		Pyrogenic	Soot	CNT Fullerenes Nanoglobules, onion-shaped nanospheres
	Inorganic	Biogenic	Oxides	Magnetite
			Metals	Ag, Au
		Geogenic	Oxides	Fe-oxides
			Clays	Allophane
		Atmospheric	Aerosols	Sea salt
Anthropogenic (manufactured, engineered)	C-containing	By-product	Combustion by-products	CNT Nanoglobules, onion-shaped nanospheres
		Engineered	Soot	Carbon Black Fullerenes Functionalized CNT, fullferenes
			Polymeric NP	Polyeth-yleneglycol (PEG) NP

		Formation		Examples
	Inorganic	By-product	Combustion by-products	Platinum group metals
		Engineered	Oxides	TiO_2, SiO_2
			Metals	Ag, iron
			Salts	Metal-phosphates
			Aluminosili-cates	Zeolites, clays, ceramics

References

Banfield, J.F., and Zhang, H. (2001) Nanoparticles in the Environment. In "Nanoparticles and the Environment" (J.F. Banfield and A. Navrotsky, Eds.), pp. 1–58. Mineralogical Society of America, Washington, DC Chapter 1

EPA. (2005) Nanotechnology White Paper The U.S. Environmental Protection Agency, Science Policy Council, Washington, D.C http://www.epa.gov/osa/pdfs/EPA_nanotechnology_white_paper_external_review_draft_12-02-2005.pdf downloaded July, 2008

Hochella, M.F., Lower, S.K., Maurice, P.A., Penn, R.L., Sahai, N., Sparks, D.L., and Twining, B.S. (2008) Nano minerals, mineral nanoparticles, and earth chemistry. *Science* **21**: 631–1635

Nowak, B and Buchelli, T.D. (2007) Occurrence, behavior and effects of nanoparticles in the environment. *Environmental Pollution* **150**: 5–22

Chapter 2

Synthesis of Nanomaterials

Solid state materials can exists in various forms namely:

1. Single crystals (preferred for authentic characterization of structure and physical properties)
2. Polycrystalline powders that are agglomerates of large number of crystallites and are used for characterization when single crystal cannot be easily obtained
3. Nanocrystalline or nanostructured materials that posses large surface area
4. Thin films that can be grown in the best crystalline form for many applications
5. Amorphous systems that have no long-range translational order

The properties of nanoparticles depend largely on synthesis procedures adopted. Some of well-known methods for synthesis of these materials are, chemical vapour deposition, sol-gel processes and hydro-thermal synthesis.

Sl. No.	Top-down method	Nano materials produced
1.	Arc discharge method	Fullerenes and carbon nanotubes (CNTs)
2.	Laser ablation	Board range of nanoparticles including CNTs
3.	Ball milling	Composites and mixtures of elemental powders
4.	Inert gas condensation	Oxides, alloys and semiconductors

Sl. No	Bottom up approach	Nano materials produced
1.	Homogeneous nucleation	Noble metal particles, gold nanoparticles
2.	Chemical vapour deposition (CVD)	Widely used to produce CNTs, fullerenes and boron nanotubes
3.	Molecular beam epitaxy (MBE)	Compound semiconductors, thin flims
4.	Sol-gel synthesis	Colloidal and oxide nanoparticles
5.	Hydrothermal synthesis	Elemental nanoparticles mostly oxides
6.	Microwave method	Various oxide nanoparticles including TiO_2

Top-down Approach

Using top-down approach we can create microstructures out of nanostructures, whereas bottom-up approach is the self-assembly of atoms or molecules into nanostructures.

Crushing
Grinding
High-energy ball milling
(Mechanical attrition)
Photolithography
Extreme UV lithography
X-ray lithography

Ball Milling

One of the top-down routes is the mechano synthesis or mechanical methods that is capable of producing nanoscale particles by mechanical attrition. The kinetic energy of grinding medium [either stainless steel or tungsten carbide (WC) ball bearing] is transferred to coarse (micrometer size) grained metals, alloys, ceramics, *etc.* whose nanoparticles are desired. The sample is laced in a canister that is made to rotate at high revolution per minute. The drum or canister rotates at a very high speed simultaneously imparting its kinetic energy to the material through the grinding medium. This is done under controlled atmosphere conditions to prevent oxidation. During this ball milling process, severe plastic deformation of the sample material leads to the formation of defects and dislocations. These defects and dislocations are subjected to mechanical deformation, shear and starin condition thus forming nanoparticles. These nanoparticles have wide particle size distribution.

When a particle undergoes deformation due to shear and strain, Griffith's theory describes the particle fracture as

$$\sigma_F = \frac{\gamma E}{c}.$$

σ_F is the stress at which crack propagation leads to catastrophic failure, γ is the surface energy of particles (J m^{-2}), E I s Young's modulus and c is the length of the crack .The tipping point is reached when the stress equals the strength of cohesion between atoms. For smaller particles, agglomeration predominates due to enhanced surface energy. When a balance is achieved among the stress, increased resistance to fracture, increased agglomeration and energy expended in milling, the minimum size of nanoparticles at the particular condition is achieved. One of the drawback of this method is that contamination can occur at certain particle size (which is material dependent).

Microfabrication

Microfabrication is a top-down technique utilizing the following process in a sequential fashion:

1. Film deposition-CVD and PVD
2. Photolithography-optical exposure and deposition of photoresist
3. Etching-Aqueous and/or plasma

Lithography

Lithography technique is one of the most important top-down synthesis. One of the various types of lithography technique is the optical lithography that employs the visible and UV radiation to transfer a pattern onto a substrate. Other lithoghapy technique include DUV (deep UV radiation) lithography, contact printing, proximity printing and projection printing. There is also particle beam lithography where particles do not undergo diffraction, and scattering is minimal. In this case, higher resolution can be achieved than optical, X-ray and electron-beam methods. Some techniques, such as extreme UV lithography (EUVL), use radiation with wavelength 11–14 nm, hence in such cases, smaller features (5 nm) can be achieved. However, this technique can only be performed at high vacuum. X-ray lithography (XRL) uses X-rays produced by a synchrotron source.

Bottom-up Approach

Bottom-up approach usually deals with the self-assembly of atoms or molecules into nanostructures. Bottom-up techniques lead to molecular electronics where molecules can be designed with specific electronic function. This technique help in the design of molecules for self-assembly into supramolecular structures, and also to connect molecules to the macroscopic world. Bottom-up approaches includes man-made synthesis of nanotubes and nanoparticles. Other examples of bottom-up methods are powder/aerosol compaction and chemical synthesis. Nature adopts the bottom-up approach for the growth of cells and living beings, whereas chemistry and biology help them assemble and control growth process.

Bottom-up Approach

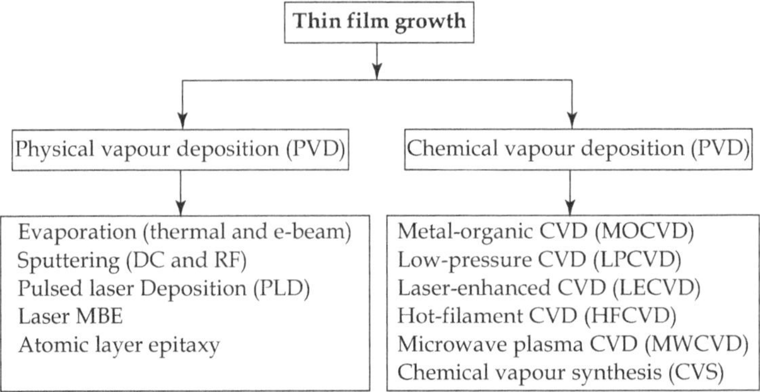

In order to eliminate interparticle interactions, nanoparticles often need to be isolated from one another by immobilization on a substrate surface or in bulk of a stabilizing inert matrix.

Physical methods of synthesis: Involves high energy treatment zin solid or gas phase

Chemical methods: In solutions at moderate temperatures

Biological methods: Involving biological sources

Various factors that govern the growth of nanoparticles are:

- Kinetics of nucleation and growth, Concentration of stabilizing agents, Structure of stabilizing agents, pH of the medium and temperature

Methods of nanoparticle synthesis should be Cost effective, monodispersity, large scale production, single step.

What are the Differences Between PVD and CVD?

PVD uses physical process. In PVD, a pure source material is first evaporated which then condenses on the substrate material eventually leading to the fabrication of thin film of the material. CVD mainly uses chemical process. Here the source material is mixed with a volatile precursor that acts as a carrier which is then injected into chamber. The chamber contains the substrate and the material get deposited on this substrate. The precursor decomposes with the help of heat, plasma or other processes and forms a thin layer of the source material on the substrate.

Chapter 3

Chemical Synthesis of Nanoparticles

Synthesis of Zn Oxide Nanoparticle

Aim

To synthesize Zn oxide nanoparticle

Introduction

Zinc oxide nanoparticle are effective anti-bacterial and anti-odur agent. They can be synthesized by number of techniques including hydrothermal methods. Their synthesis using solvothermal process is described, (Aneesh et al., 2007).

Reagents and Apparatus

- Zinc Acetate Dihydrate [(CH_3COO_2 Zn. $2H_2O$]
- Sodium hydroxide (KOH)
- Methanol
- Autoclave
- Distilled water
- Round Bottom Flask
- Magnetic Stirrer

Procedure

- Prepare stock solutions of Zn $(CH_3COO)_2.2H_2O$ (0.1 *M*) in 50 mL methanol under stirring.

- Add to this stock solution, 25 mL of NaOH (varying from 0.2 *M* to 0.5 *M*) solution prepared in methanol under continuous stirring in order to get the pH value of reactants between 8 and 11.
- Transfer these solutions into teflon lined sealed stainless steel autoclaves and maintain at various temperature in the range of 100–200 °C for 6 and 12 h under autogenous pressure.
- Allow these system to cool naturally to room temperature. Filter through Whatman No 1 filter paper.

References

Aneesh, P.M., Vanaja K.A. and Jayaraj, M.K. (2007) Synthesis of ZnO nanoparticles by hydrothermal method. Nanophotonic Materials IV, edited by Zeno Gaburro, Stefano Cabrini, Proc. of SPIE Vol. 6639, 66390J, (2007) 0277-786X/07/$18 doi: 10.1117/12.730364.

Chapter 4

Synthesis of Iron (III) Oxide Nanoparticles

Aim

To synthesize iron (III) oxide (Fe_2O_3) nanoparticles

Introduction

Iron oxide nanoparticles have a number of applications in catalysis, sensors and high sensitive biomolecular magnetic resonance imaging (MRI) for medical diagnosis and therapeutics. They can be synthesize by the bottom-up approaches including the sol-gel technique.

Reagents and Apparatus

- Sodium hydroxide (NaOH)
- Iron Chloride ($FeCl_3$)
- Sodium hexametaphosphate ($Na_6P_6O_{18}$)
- Hydrochloric Acid (HCl)
- Distilled water
- Round Bottom Flask
- Magnetic Stirrer

Procedure

- Prepare 0.6 *M* NaOH solution in distilled water and add 0.01 *M* Solution of $FeCl_3$ and stir it for 30 min at 100 °C. A precipitate of $Fe(OH)_3$ will be obtained (Shah and Shah, 2013).

- Wash the precipitate with distilled water several times till pH becomes 8.8.
- To the precipitate add 1 ml of 1 *M* HCl and 10 ml of 0.1 *M* solution of $Na_6P_6O_{18}$ in distilled water, stir and heat upto 100 °C.
- Filter the precipitate of Fe_2O_3 formed wash with distilled water and dry in air.

Results

The nanoparticles of Fe_2O_3 are obtained

References

Shah, M.A and Shah, K.A. (2013) Nanotechnology: The Science of Small The Wiley India Pvt. Ltd, pp 167.

Chapter 5

Synthesis of Nano (Silver) Ag

Aim

To synthesize silver (Ag) nanoparticle.

Introduction

The Ag nanoparticles easily interact with other particles and increases their antibacterial efficiency. The Ag nanoparticles can be synthesized using various methods: biological, chemical electrochemical, γ-radiation, photochemical, laser ablation *etc* (Udapudi et al., 2012). The most popular preparation of Ag colloids is chemical reduction of Ag salts by sodium borohydride or sodium citrate (Udapudi et al., 2012).

Reagents and Apparatus

- 0.001 *M* Silver nitrate solution
- 1% tri-sodium citrate
- Magnetic stirrer-hot plate
- Uv-Vis spectrophotometer

Procedure

- Heat 50 ml of 0.001 *M* Silver nitrate in 500 mL glass beaker to boiling using hot plate magnetic stirrer.
- Add to this solution 5 mL of 1% tri-sodium citrate drop by drop. Mix solution vigorously.

- Continue heating until color change is evident (yellowish brown).
- Remove the solution from the heating element and stir until cool to room temperature.

Reaction mechanisms

$$4Ag^+ + C_6H_5O_7Na_3 + 2H_2O \rightarrow 4Ag^0 + C_6H_5O_7H_3 + 3Na^+ + H^+ + O_2$$

Results

Colloidal Ag nanoparticles produced.

Utility

One gram of silver nanoparticles is all that is required to give antibacterial properties to hundreds of square meters of substrate material.

References

Udapudi, B., Naik, P., Sabiha, S.T., Sharma, R. and Balgi, S. (2012) synthesis and characterization of silver nanoparticles. *International Journal of Pharmacy and Biological Sciences* **2**:10–14.

Chapter 6

Synthesis of Chitosan/Nanosilica Coating Solution

Aim

To synthesize chitosan/nanosilica coating solution for fruit coating aiming at increasing shelf life.

Introduction

Application of edible coatings is promising to improve the quality and extend shelf life of fruits and vegetables (Lin *et al.*, 2011). Chitosan, a versatile biopolymer derived from deacetylation of chitin had been widely applied in the fresh-keep field owing to its good biocompatibility, biodegradability, antibacterial activity and capacities to form film (Lin *et al.*, 2011).

Reagents and Apparatus

- Martin Rose Bengal Agar medium
- High molecular weight chitosan, with a deacetylation degree of 75.6%
- 98% glacial acetic acid
- Tetraorthosilicate (TEOS)

Procedure

- Prepare chitosan/nano-silica hybrid film according to Yeh *et al.* (2007) with minor modification. Disperse chitosan

(2.0%, w/v) coating solution in an aqueous solution of glacial acetic acid (0.5%, v/v).

- Stir the solution with a magnetic stirrer for 48 h at room temperature to form a homogeneous mixture. Take required amount of this solution in a 500 mL bottle and add measured amount of TEOS to it.
- Stir the solution repeatedly with a magnetic stirrer until the mixture solution became clear. Add stoichiometric amount of an equal mixture of ethanol and water to the solution and allow to stir for 20 h at room temperature.
- Adjust water to TEOS ratio to 1:4. Adjust the pH of the solution was to 5.6 with 1 *M* NaOH.
- Vacuum the 500 mL final solution for 2 h to remove trapped air bubbles during the mixing.

Reaction Mechanisms

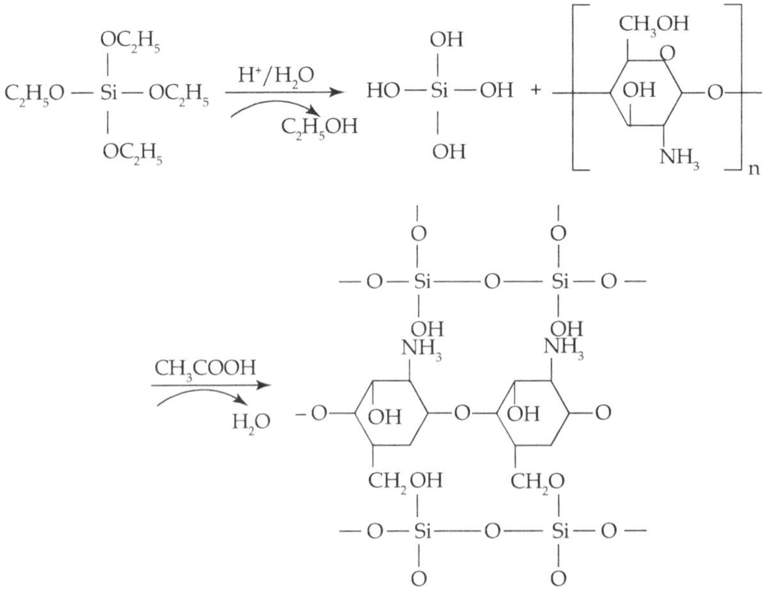

Fig. Schematic representative for preparation of the silica-coated chitosan particle using molecular level interaction between silanol and chitosan polymer

Utility

The chitosan/ nano-silica composite could provide a promising alternative to improve the preservation qualities of fresh longan fruits during extended storage. In addition, natural and artificial chitosan-inorganic composites will offer intriguing opportunities for research in food nano-science and also provide new chances for innovation with tremendous possibilities in bringing solutions for the fruit storage and bioprocessing industry.

References

Lin, B.F., Du, Y.M., Liang, X.Q., Wang, X.Y., Wang, X.H., Yang, and J.H. (2011) Effect ofchitosan coating on respiratory behavior and quality of stored litchi under ambient temperature. *Journal of Food Engineering* **102**: 94–99.

Yeh, J.T., Che, C.L. and Huang, K.S. (2007) Synthesis and properties of chitosan/SiO_2hybrid materials. *Materials Letters* **61**: 1292–1295.

Chapter 7

Equilibrium Water Absorbency of Hydrogel

Introduction

Hydrogels are loosely cross-linked, highly hydrophilic, organic polymers. They can absorb and retain aqueous fluids up to 500 times of their own weight (Singh *et al.*, 2011). In the field application, such absorbents exhibited a limited swelling capacity within the soil matrix.

Experiment

- Immerse a composite sample weighing 0.1 g (passed through 2 mm sieve) in the excess of distilled water (pH 7.0, EC 0.001 mhos/cm) in triplicate and keep it in room temperature (@ 25 °C) separately until equilibrium was attained.
- Free water was filtered through a nylon sieve (200 mesh size), gel allowed to drain on sieve for 10 min, and finally weighed.
- The water absorbency (Q_{H_2O}) was calculated using the following equation:

  $$Q_{H_2O}\ (\text{g g}^{-1}) = (w_2 - w_1)/w_1$$

  where w_1 is the weight of dry gel and w_2 is the weight of swollen gel. Q_{H_2O} was calculated as grams of water per gram of dry sample.

Reference

Singh, A., Sarkar, D.J., Singh, A.K., Parsad, R., Kumar, A., Parmar, B.S. and Singh, B.S. (2011) Studies on novel nanosuperabsorbent composites: swelling behavior in different environments and effect on water absorption and retention properties of sandy loam soil and soil-less medium. *Journal of Applied Polymer Science* **120:** 1448–1458.

Chapter 8

Zerovalent (nZVI) Nanostructures Application in Heavy Metal Decontamination

Zerovalent (nZVI) Ionic Compounds

Although Ag^+ contains a filled d^{10} shell, the structures of Ag compounds can in many instances be understood only if one assumes the existence of weak attractive $Ag^+ - Ag^+$ interactions. Similar attractive forces have been shown to be responsible for the formation of gold cluster compounds. In Ag^+ solid state compounds the unoccupied 5s and 5p orbitals give rise to empty bands which may be partially occupied thus resulting in subvalent Ag compounds; examples are Ag_3O and Ag_2F. The compound Ag_5SiO_4 prepared from Ag and $Si0_2$ at 350 °C under a high pressure of oxygen, contains $[Ag_6]^{4+}$ clusters (average oxidation state 2/3) and should be formulated at $[Ag_6]^{4+}$ $(Ag^+)_4$ $(SiO_4^{4-})_2$.

Ag and Au forms numerous cluster compounds in which the metals are zero valent or subvalent. Examples are Ag_6 Fe_3 $(CO)_{12}$ $\{HC(PPh_2)_3\}$ and paramagnetic cluster ions $[Ag_{13}$ Fe_8 $(CO)_3^{2}]^{4-}$.

Zerovalent (nZVI) Particle Structure

Bare nZVI particles are typically less than 100 nm in diameter. In aquesos solution, all nZVI. Fe reacts with water and oxygen to form an outer (Fe) hydroxide layer. As a result all nZVI have core shell structure. The thin and distorted oxide layer allows electron transfer from the metal (1) directly through defects such as pits or pinholes, (2) Indirectly *via* conduction band, impurity, localized band and (3) from sorbed or structural Fe^{2+} thus sustain the capacity of the particle for reduction of contaminants. The outer hydroxide layer may act as efficient adsorbent for various contaminates including metals (O'Carroll *et al.*, 2013).

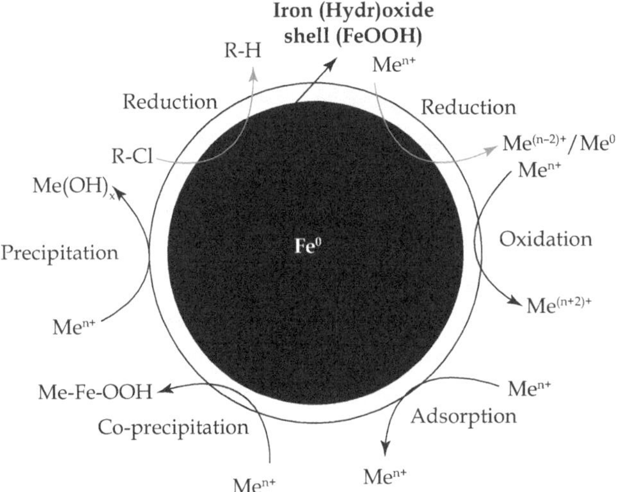

Fig: Core shell structure of Zero-valent iron

Synthesis of Zero Valent Iron

The nZVI can be synthesized *via* number of methods including the sonochemical methods, the electrochemical methods, the gas phase reduction method and the liquid phase reduction method. Gas phase and liquid phase reduction methods are most commonly used for synthesizing nZVI for remediation purpose.

Reaction of Heavy Metals with Nanomaterials

The transformation, solubility, mobility and consequently toxicity of heavy metals in the environment are governed by redox reactions, precipitation/dissolution and adsorption/desorption phenomenon. Water treatment strategies for removal of metal contaminants typically involve manipulating these mechanisms to control availability and toxicity to biota. Chromium at higher oxidation state (Cr^{6+}) is very toxic but at lower oxidation state (Cr^{3+}) is essential relatively non toxic, but toxic at high concentration.

The specific reaction mechanisms involved in treatment of heavy metal with nZVI depend on the standard redox potential of metal contaminant (E^0). Metals that have E^0 more negative than, or similar to, that of Fe^0 (eg Cd, Zn) are removed purely by adsorption to the iron (hydr)oxide shell. Metals with E^0 more positive than Fe^0 (eg Cr, As, Cu, U and Se) are preferentially removed by reduction and precipitation. Metals with slightly more positive E^0 than Fe^0 (Pb and Ni) are removed by both reduction and precipitation. Oxidation-reduction and coprecipitation are other reaction mechanisms.

Aqueous solution	Half-cell reaction	Eo(V)
Chromium (Cr)	$CrO_4^{2-} + 8H^+ + 3e^{-1} \leftrightarrow Cr^{3+} + 4H_2O$	1.51
Chromium (Cr)	$Cr_2O_7^{2-} + 14H^+ + 6e^{-1} \leftrightarrow 2\,Cr^{3+} + 7H_2O$	1.36
Arsenic (As^{V})	$H_3ASO_4 + 2H^+ + 2e^{-1} \leftrightarrow HAsO_2 + 4H_2O$	0.56
Arsenic (As^{III})	$H_3ASO_3 + 3H^+ + 3e^{-1} \leftrightarrow As + 3H_2O$	0.24
Cadmium (Cd)	$Cd^{2+} + 2e^{-1} \leftrightarrow Cd$	0.40
Nickel (Ni)	$Ni^{+2} + e^{-1} \leftrightarrow Ni$	–0.25
Lead (Pb)	$Pb^{2+} + 2e^{-1} \leftrightarrow Pb$	–0.13
Iron (Fe)	$Fe^{2+} + 2e^{-1} \leftrightarrow Fe$	–0.44
Zinc (Zn)	$Zn^{2+} + 2e^{-1} \leftrightarrow Zn$	–0.76

Characterization

A variety of surface analysis techniques such as X-ray photoelectron spectroscopy (XPS), extended X Ray absorption spectroscopy (EXAFS), X-ray absorption near edge structure spectroscopy (XANES), X-ray diffraction (XRD) and scanning transmission electron microscopy (STEM) with energy dispersive X-ray (EDX) can be used to evaluate metal removal mechanism associated with nZVI.

Categorization of metal-nZVI interactions

S. No.	Reaction mechanism	Examples
1.	Reduction	Cr, As, Cu, Pb, Ni, Se, Hg, Ag
2.	Adsorption	Cr, As, Pb, Ni, Se, Cco, Cd
3.	Oxidation-reduction	As, Se, Pb, U
4.	Co-precipitation	Cr, As, Ni, Se
5.	Precipitation	Cu, Pb, Cd, Co, Zn

Reference

O'Carroll, D., Sleep, B., Krol, M., Boparai, H. and Kochur, C. (2013) Nanoscale Zero Valent Iron and bimetallic particles for contaminated site remediation. *Advances in Water Resources* **51**: 104–122.

Chapter 9

Nanocomposite Technology

Polymer Clay Nanocomposites

Polymer clay nanocomposites has been used as to affect permeability of the maternal to gases and vapours. Majority of papers on nanocomposites have been focused on the use of smectitic type clays as nanoparticles. They are a group of swelling type of clay minerals including montmorilonite, nontronite, saponite, sauconite and hectorite.

A number of polymer-layered silicates (PLS) preparation methods has been reported in literature. The three most common methods to synthesize PLS nanocomposite are intercalation of suitable monomer and subsequent *in situ* polymerization, intercalation of polymer from solution and polymer melt intercalation.

In general, layered silicates have layer thickness on the order of 1 nm and a very high aspect ratio (*e.g.* 10–1000). A few weight percent of layered silicates that are properly dispersed throughout the polymer matrix thus create much higher surface area for polymer/filler interaction as compared to conventional composites. Depending on the strength of interfacial interactions between the polymer matrix and layered silicate (modified or not), three different types of PLS nanocomposites are thermodynamically achievable (Fig. 1 and 2) as outlined by Sinha Ray and Okamoto (2003). They are as follows:

a. Intercalated nanocomposites: In intercalated nanocomposites, the insertion of a polymer matrix into the layered silicate structure occurs in a crystallographically regular fashion, regardless of the clay to polymer ratio. Intercalated nanocomposites are normally interlayer by a few molecular layers of polymer. Properties of the composites typically resemble those of ceramic materials.
b. Flocculated nanocomposites: Conceptually this is same as intercalated nanocomposites. However, silicate layers are sometimes flocculated due to hydroxylated edge–edge interaction of the silicate layers.
c. Exfoliated nanocomposites: In an exfoliated nanocomposite, the individual clay layers are separated in a continuous polymer matrix by an average distances that depends on clay loading. Usually, the clay content of an exfoliated nanocomposite is much lower than that of an intercalated nanocomposites.

Exfoliated nanocomposites offers complete dispersion of clays into the polymer matrices and at very low clay concentration as compared to other types (Flocculated and Intercalated). Exfoliated nanocomposites may be very useful in achieving the desired barrier properties as compared to others.

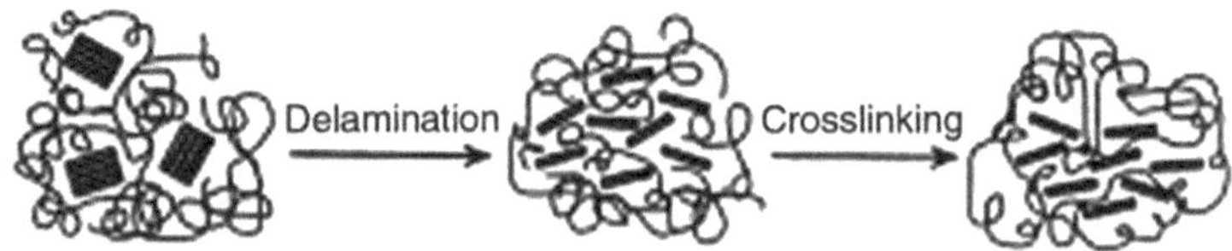

Fig. 1 Schematic diagram of nanocomposite synthesis

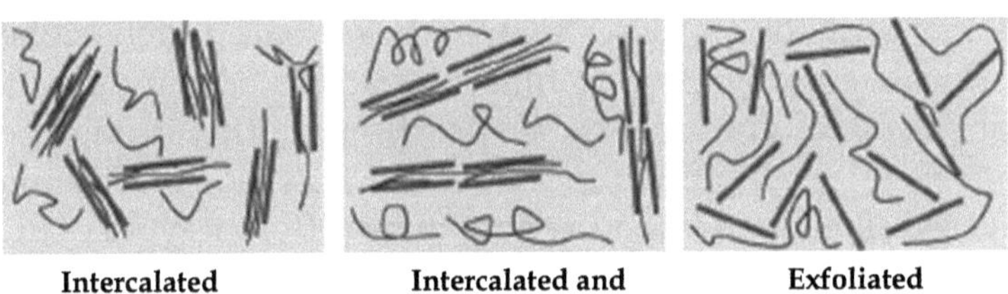

Fig. 2 Three types of nanocomposites as outlined by Sinha Ray and Okomoto, 2003

Mostly smectitic type of clay minerals has been used because they are naturally ***occurring clay minerals that are commercially available and exhibit platy morphology with high aspect ratio and substantial cataion exchange properties.***

The platelet structure of these aluminosilicates materials has proven its ability to improve the barrier properties of polymericmaterials, according to a tortuous path model in which a small amount of platelet particles significantly reduces the diffusivity of gasses through the nanocomposites. The key of nanocomposite technology is exfoliation of the clay into its individual platelets, thereby achieving its greatest barrier improvement as well as lowest haze. The advantage of this approach is based on the use of conventional cheap polymers with barrier improvement arising from dispersion of low levels of inexpensive clay minerals.

References

Sinha-Ray, S. and Okamoto, M. (2003) Polymer/layered silicate nanocomposites: A review from preparation to processing. *Progress in Polymer Science* **28:** 1539–1641.

Chapter 10

Green Synthesis of Nanoparticles

Biological synthesis of Zinc Oxide (ZnO) nanoparticles

Aim

To synthesize nano ZnO particles using fungal isolates.

Introduction

The ZnO nanoparticle are effective anti-bacterial and anti-odur agent. Biological synthesis of ZnO nanoparticle is a new approach for environmentally benign protocol in context to green nanotechnology, (Raliya and Tarafdar, 2013).

Reagents and Apparatus

- Martin Rose Bengal agar medium
- BOD incubator
- Potato dextrose agar (PDA) media
- Rotary shaker
- Biosafety cabinet
- Zinc nitrate ($ZnNO_3$)

Procedure

Isolation of Fungi

- Isolate the fungi, *Aspergillus fumigatus* from soil by plating the inoculum on Martin Rose Bengal Agar medium

(Hi-Media, India, pH 7.2) after serial dilutions of pooled soil sample.

- Inhibit bacterial contamination by supplementing the medium with chloramphenicol (Sigma-Aldrich, St. Louis, USA) at a concentration of 10 μg mL^{-1} after autoclaving.
- Incubate inoculated plates at 28 °C for 72 h in BOD incubator. Pick up individual fungal colonies and further purify by sub-culturing on PDA media (Hi-Media, India).
- Perform preliminary identification of fungal isolates was on the basis of morphological characteristics.

Biosynthesis of ZnO Nanoparticle

- Grow the fungi *Aspergillus fumigatus* in 250-mL Erlenmeyer flask containing 100-mL broth medium, composed of 0.3% malt extract, 1% sucrose, 0.3 % yeast extract, and 0.5% peptone.
- Adjust the pH of the medium to 5.8, and grow the culture with continuous shaking on a rotary shaker (150 rpm) at 28 °C for 72 h.
- Separate out fungal balls of mycelia after complete incubation, from the culture broth by filtration using Whatman No. 1 filter paper under biosafety cabinet and then wash the fungal mycelia thrice with sterile double-distilled water.
- Re-suspend the harvested fungal mycelia (10-g wet weight) in 100-mL sterile Milli-Q-water in 250-mL Erlenmeyer flask and again put into a rotary shaker (150 rpm) at 28 °C for 48 h.
- Obtain the cell-free filtrate after incubation by separating the fungal biomass using 0.45-μm membrane filter.
- Using cell-free filtrate, prepare salt solution of zinc nitrate ($ZnNO_3$) with final concentration of 0.1 m*M* in Erlenmeyer flasks, which is the optimum salt concentration for the synthesis of monodispersed ZnO.
- Put the entire mixture into rotary shaker at 28 °C at 150 rpm. Allow the reaction to carry out for a period of 72 h.

- Periodically characterize the bio transformed product for particle size.

Characterization

- **Particle size distribution and colloidal stability:** Monitor the particle size distribution and zeta potential of ZnO nanoparticles using dynamic light scattering (DLS) measurements by measuring the rate of fluctuations in the laser light intensity scattered by particles as they diffuses through solvent.
- **Size and shape of nano ZnO particles** : Carry out transmission electron microscopif (TEM) measurements for the confirmation of size and shape using drop coating method in which place a drop of solution containing nanoparticles on the carbon-coated copper grids and keep under vacuum desiccation for overnight before loading them onto a specimen holder.
- **Topology and surface observation:** Determine topology and surface observations by scanning electron microscopy (SEM) as it offer better resolution and depth of field than optical microscope. Examine the crystal structure and size by X-ray diffraction (XRD) analysis.

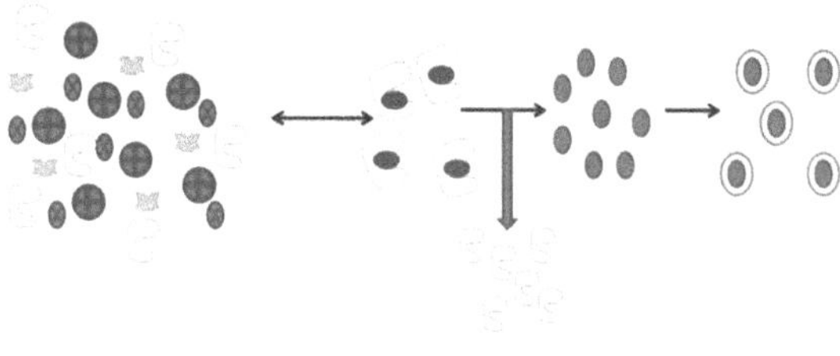

Fungal extracellular enzyme
Zinc ion
Nitrate ion
Water molecule
ZnO nanoparticle
Protein encapsulated ZnO nanoparticle

Fig. Hypothetical mechanisms of Biological synthesis of Nano ZnO particles

References

Raliya, R. and Tarafdar, J. C. (2012) ZnO nanoparticle biosynthesis and its effect on phosphorous-mobilizing enzyme secretion and gum contents in clusterbean (*Cyamopsis tetragonoloba* L.) *Agricultural Research* **2(1):** 48–57.

Chapter 11

Biological Synthesis of Phosphorus Nanoparticles

Aim

To synthesize phosphorus (P) nanoparticles from tricalcium phosphate using fungal mycelium.

Introduction

Green nanotechnology includes a range of processes that reduce or eliminate toxic substances to restore environment. Microorganisms like fungi and bacteria having naturally bestowed property of reducing and oxidizing nanoparticles, are used for synthesis.

Reagents and Apparatus

- Martin Rose Bengal agar medium
- BOD incubator
- Potato dextrose agar (PDA) media
- Rotary shaker
- Tri-calcium phosphate ($Ca_3P_2O_8$)
- Biosafety cabinet

Procedure

Isolation of fungi Isolation of Fungi

- Isolate *Aspergillius tubingensis* TFR-5 from soil using standard protocols including serial dilution followed by spread plate method using Rose Bengal Agar medium (Hi-Media, India), which is especially used for cultivation of fungi.
- Incubate the inoculated plates at 28 °C for 72 hours in BOD incubator. Pick up individual fungal coloni and further purify by subculturing on Potato Dextrose Agar (PDA) media.
- Maintain stock culture by sub culturing at a regular time interval.
- Preserve the slant −20 °C after growing at pH 5.8 and 28 °C for 72 hours, the slants at.
- Prepare on the fresh PDA slant from an actively growing stock culture and after 72 hours of incubation at 28 °C., useas starting material for Phosphorous nanoparticle synthesis.

Biosynthesis of Phosphorus Nanoparticles

- Grow the fungus *Aspergillius tubingensis* TFR-5 in 250 mL erlenmeyer flask containing 100 mL broth medium, composed of 0.3% malt extract, 1% sucrose, 0.3% yeast extract, and 0.5% peptone. Grows the fungus with continuous shaking on a rotary shaker (150 rpm) at 28 °C for 72 hours after adjusting the pH of medium to 5.8.
- Separate out fungal balls of mycelia after complete incubation from the culture broth by filtration process using whatman No. 1 filter paper under biosafety cabinet and wash the fungal mycelia thrice with sterile double distilled water.
- Re-suspend the harvested fungal mycelia (10 g wet weight) in 100 mL sterile Milli-Q-water in 250 mL Erlenmeyer flask and again put into a shaker (150 rpm) at 28 °C for 48 hours.

- Obtain cell free filtrate after incubation by separating the fungal biomass using 0.45 μ size membrane filter .
- Prepare 0.1 m**M** salt solution of tri calcium phosphate ($Ca_3P_2O_8$) using cell free filtrate of in Erlenmeyer flasks (Raliya *et al*, 2012).
- Put the entire mixture into shaker (150 rpm) at 28 °C and allow the reaction to carry out for a period of 72 hours. Use the bio-transformed product periodically for characterization based on particle size.

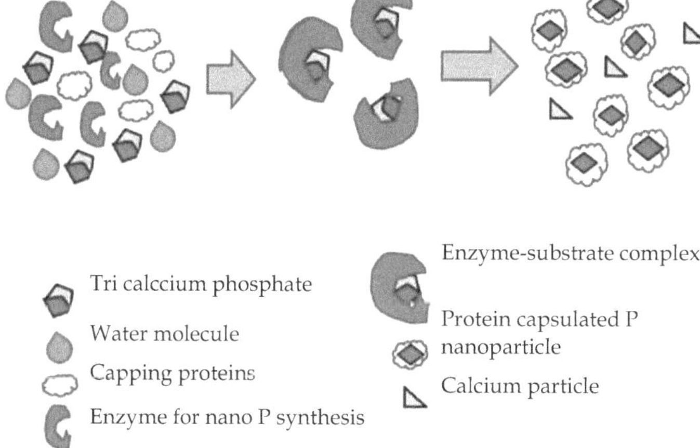

Fig. Hypothetical mechanism of nanophosphous synthesis

Characterization

- **Size and shape of nanoparticles**: Measure the particle size distribution of P nanoparticles using DLS measurements .
- **Size and shape of biosynthesized P nanoparticles**: Confirm size and shape using TEM measurements using drop coating method.
- **Surface topography** : Use atomic force microscopy (AFM) for studying near-atomic-resolution with 3D surface topography, (Veeco di CP-II scanning probe microscope in atomic force, tapping mode with closed loop 5 × 5 micron scanner).
- **Elemental composition analysis**: Perform elemental composition analysis using electron dispersive X-ray

spectroscopy (EDX). Prepare samples on a carbon coated copper grids and keep under vacuum desiccation for three hours before loading them onto a specimen holder. Determine the elemental composition and purity of the sample by atom.

Utility

The great interest lies in the possibility of using nanoparticles in agricultural, biomedical and electronic applications with recent implications in optical devices, biolabeling, drug delivery, catalysis, nanotherapeutics and nanodiagnostics.

References

Raliya, R., Tarafdar, J.C. and Rathore, I. (2012) Microbial synthesis of phosphorous nanoparticle from Tri-calcium phosphate using *Aspergillus tubingensis* TFR-5. *Journal of Bionanoscience* **6**: 1–6.

Chapter 12

Biological Synthesis of Gold Nanoparticles

Aim

To synthesize gold (Au) nanoparticle using fungal mycelium.

Introduction

Biosynthesis of nanoparticles is a kind of bottom up approach where the main reaction occurring is reduction/oxidation. The fungal enzymes are usually responsible for reduction of metal compounds into their respective nanoparticles.

Reagents and Apparatus

- Martin Rose Bengal agar medium
- Chloramphenicol
- BOD incubator
- Potato dextrose agar (PDA) media
- Rotary shaker
- Chloroauric acid ($HAuCl_4$)
- Biosafety cabinet

Procedure

Isolation of fungi

- Isolate the fungi, *Rhizoctonia bataticola* TFR-6 (NCBI GenBank Accession No. JQ675307) from soil through serial dilution and pate the inoculum on Martin Rose Bengal Agar medium (HiMedia, India, pH 7.2) after serial dilutions of collective soil sample.

- Add 10 µg mL^{-1} Chloramphenicol during autoclaving for preventing bacterial growth contamination. Incubate inoculated plates at 28 °C for 72 h in BOD incubator .
- Pick up fungal colonies and further purify by sub-culturing on PDA media.
- Identify fungal isolates on the basis of morphological characteristics.

Biosynthesis of Gold Nanoparticles

- Grow the fungi, *Rhizoctonia bataticola* TFR-6 in 250 mL Erlenmeyer flask containing 100 mL potato dextrose broth medium.
- Adjust the pH of the medium to 5 and shake continuously on a rotary shaker (150 rpm) at 28 °C for 72 h.
- Isolate the fungal balls of mycelia from the culture broth by filteration using Whatman filter paper no. 1 under biosafety cabinet and wash the fungal mycelia thrice with sterile double distilled water.
- Re-suspend the harvested fungal mycelia (20 g wet weight) in 100 mL sterile Milli-Q-water in 250 mL Erlenmeyer flask and again put into a rotary shaker (150 rpm) at 28 °C for 12 h.
- Obtain the cell free filtrate by separating the fungal biomass using 0.45 µ size membrane filter after incubation.
- Prepare 0.1 m*M* $HAuCl_4$solution using cell free filtrate in Erlenmeyer flasks for synthesis of monodisperse gold nanoparticles (Raliya and Tarafdar, 2013).
- Keep the entire mixture on rotary shaker at 28 °C at 150 rpm. Allow the reaction to carry out for a period of 4 h.
- Periodically characterize the bio transformed product using particle size analyzer.

Characterization

- **Particle size analysis**: Monitor the particle size distribution of gold nanoparticles using DLS measurements.
- **Size and shape of nanoparticles:** Confirm size and shape of biosynthesized nanoparticles TEM measurements using drop coating method.
- **Surface topography:** Use AFM for studying near-atomic-resolution with 3D surface topography.
- **Elemental composition analysis:** Perform elemental composition analysis using EDX.

References

Raliya, R. and Tarafdar, J.C. (2013) Biosynthesis of gold nanoparticles using *Rhizoctonia Bataticola* TFR-6. *Advanced Science, Engineering and Medicine*. **5**: 1–4.

Chapter 13

Phytogenic Synthesis of Silver Nanoparticles

Aim

Phytosynthesis of silver (Ag) nanoparticles using *Cassia roxburghii* aqueous leaf extract.

Introduction

The Ag nanoparticles constantly exhibit adverse effects on microbes such as inhibition and inactivation (Nasrollahi *et al.*, 2011). The Ag nanoparticles synthesized by using various biological sources such as herbs, plants and biological organisms display excellent properties in various fields i.e., non-linear optics and intercalation materials for electrical batteries as optical receptors, catalysts in chemical reactions biolabelling and as antibacterials (Sanghi and Verma, 2009).

Reagents and Apparatus

1. *Cassia roxburghii* leaf
2. Extraction assembly
3. $AgNO_3$
4. BOD incubator shaker with temperature control

Procedure

Preparation of leaf extract

Take 4 g of *Cassia roxburghii* leaf powder and mix with 100 mL of glass distilled water.

Synthesis of Nano Ag particles

- Take 1.0 mL of the leaf aqueous extract and add to 9.0 mL of 1.0 m*M* $AgNO_3$ solution with pH maintained at 7.0 and temperature maintained at 37°C.
- The reaction should be performed in dark at room temperature for 24 h in a static condition and recorded for any change in colour.
- Analyze the synthesized Ag nanoparticles using UV-vis Spectrophotometer under 300 to 700 nm wavelength.
- Visualize and record the samples for any precipitation or agglomeration of nanoparticles.

Characterization

- Incubate the Ag nanoparticles for longer period (1-2 months) under dark at 3TC, and periodically characterize through UV-vis spectro photometry (λ_{max} – 430 nm) and High-resolution transmission electron microscopy (HRTEM).
- Place a drop of Ag nanoparticles on carbon-coated copper grids and allow to stand for two min. Remove excess solution using a blotting paper and dry at room temperature. The particles are generated predominantly spherical in shape with a diameter ranging from 10 to 30 nm.
- The hydrodynamic diameter using the diffusion coefficient of the monodispersive colloids of Ag nanoparticles and the auto correlation function should be measured using DLS at 35 nm wavelength.
- The electrophoretic mobility gives the charges which has an impact on the particle distribution. The higher the zeta potential, the higher will be the electric charge on the surface of the particles. Using water as dispersant the zeta potential can be found to be −18.3 mV for Ag nanoparticles. The intensity of negative charge indicates stability.

Utility

The growth of phytopathogenic fungi are controlled by synthetic fungicides; however, the uses of these are increasingly restricted due to their harmful effects on human health and the environment. The phytopathogenic fungi such as *R. solani and F. oxysporum* are known to cause the disease and attribute to a loss of 5–60% yield . The increasing demand of production and regulations on the use of agrochemicals and the emergence of resistant pathogen to the products employed, justifies the search for novel active molecules and new control strategies.

References

Nasrollahi, A., Pourshamsian, K. and Mansourkiaee, P. (2011) Antifungal activity of silver nanoparticles on some of fungi. *International Journal of Nano Dimension* **1** : 233–339.

Sanghi, R. and Verma, P. (2009) Biomimetic synthesis and characterization of protein capped silver nanoparticles. *Bioresource Technolology* **100**: 501–504.

Chapter 14

Green Synthesis of Zero Valent Iron (Fe) (nZVI)

Aim

To synthesize Zero valent (Fe) (nZVIs) iron using tea leaf extract.

Introduction

In this approach extracts of natural products (in most cases green tea leaves) with high antioxidant capacities are used. The compounds present in these extracts react with Fe (III) in solution to form nZVIs (Nadagouda *et al.*, 2010).

Reagents and Apparatus

- Ethanol (96%)
- Folin-Ciocalteu reagent
- Iron(II) sulfate heptahydrate
- 2,4,6-Tris(2-pyridyl)-s-triazine (Sigma-Aldrich)
- Sodium carbonate (99.8%)
- Gallic acid (98.0%)
- Sodium acetate trihydrate (99.0%)
- Glacial acetic acid (99.7%)
- Hydrochloric acid (37%)
- Iron(III)chloride hexahydrate (99.0%)
- Type II deionized water (Millipore)

Procedure

- Pick tea leaves and mill using a normal kitchen chopper.
- Sieve chopped leave using a 4-mm sieve and use only the material with sizes 4 mm.
- Divide the materials into two parts: Use one part directly in the extraction process and pre-dry the other at 50 °C in an oven for 48 h.
- Determine the moisture content of each leaf sample.

Determination of antioxidant capacity

- Measure the antioxidant capacity of the extracts using the "ferric reducing antioxidant power" (FRAP) method (Pulido et al., 2000).
- Determine total phenolic content total phenolic content of the extracts, obtained using the optimized Procedures the Folin-Ciocalteu method (ISO, 2005).

Extraction procedure

- Weigh 1.8 g of leaf and transfer to a 100-mL Erlenmeyer flask and add 50 mL of water.
- Place the flask in a shaker bath (Stuart Scientific, SBS30) at room temperatures 80 °C for 120 min and draw samples in every 20 min.

Production and size determination of the nanoparticles

- Evaluate the capacity of the extracts to produce nZVIs by mixing 1 mL of extract and 250 µL of an Fe (III) solution (0.1 *M*) (Machado, *et al.*, 2013).
- Mount the resulting dark colored solutions on 300-mesh nickel grids and examine using a JEOL JEM 1400 Transmission Electronic Microscope (TEM; 120 kV). Determine the size of the produced nZVIs, using magnification modes of 120,000 to 500,000 .

Utility

- Zero-valent iron started to be used in permeable reactive barriers (PRBs) in the beginning of the 1990s for the treatment of water, wastewater, groundwater, soil, sediments, and gaseous streams, mostly contaminated with chlorinated contaminants (trichloroethylene and/or perchloroethylene), but also with heavy metals (chromium or lead) or organochlorine pesticides (lindane or DDT) (Huang *et al.*, 2012).

References

Machado, S., Pinto,S.L., Grosso, J.P., Nouws, H.P.A., Albergaria, J.T. and Delerue-Matos, C. (2013) Green production of zero-valent iron nanoparticles using tree leaf extracts. *Science of the Total Environment* **445–446:**1–8.

Huang J.L., Cong, X. and Gu, Q.B. (2012) Factors influencing the dechlorination of organo-chlorine pesticides in soils of a contaminated site by zero-valent iron. *Disaster Advances* **5**:105–8.

Nadagouda, M.N., Castle, A.B., Murdock, R.C., Hussain, S.M. and Varma, R.S. (2010) In vitro biocompatibility of nanoscalezerovalent iron particles (NZVI) synthesized using tea polyphenols. *Green Chemistry* **12**:114–22.

ISO (2005) International Organization for Standardization, ISO 14502-1. Determination of substances characteristics of green and black tea—part 1: content of total polyphenols in tea—colorimetric method using Folin–Ciocalteu reagent; 2005. p. 14.

Pulido, R., Bravo, L and Saura-Calixto F. (2000) Antioxidant activity of dietary polyphenols as determined by a modified ferric reducing/antioxidant power assay. *Journal of Agriculture and Food Chemistry* **48**:3396–402.

Chapter 15

Chemical Synthesis *vs* Green Synthesis

Chemical Synthesis Methods

Several methods can be used for the production of these nanoparticles, namely: (i) top-down methods (Li *et al.*, 2006) such as vacuum sputtering (Kuhn *et al.*, 2002) or the decomposition of iron penta carbonyl ($Fe(CO)5$) in organic solvents (Karlsson *et al.*, 2005); and (ii) bottom-up methods that promote the 'growth' of the nanostructures *via* chemical synthesis, for example through the reaction of iron(II) or iron(III) salts with sodium borohydride (Wang and Zhang, 1997). However, these methods present several limitations and problems; the top-down methods are generally expensive and require specific and costly equipment while the drawbacks of the bottom-up approaches are related to safety issues due to the toxicity of sodium borohydride, the production of flammable hydrogen gas during the process (Li *et al.*, 2006), and the tendency to form large agglomerates, very fast and at a high extent, and therefore have a reduced reactivity and degradation efficiency.

Green Synthesis Methods

These disadvantages created incentives to incorporate green chemistry principles, such as the choice of 'greener' solvents and reducing agents or the utilization of appropriate capping agents (Hoag *et al.*, 2009). Therefore, and in the last couple of years, a greener approach for the production of nZVI has been developed. In this approach extracts of natural products (in most cases green tea leaves with high antioxidant capacities are used. The compounds present in these extracts react with iron(III) in solution to form the nZVIs (Nadagouda *et al.*, 2010). The main advantages of this method are: (i) the lower toxicity of the used reducing agent compared with borohydride; (ii) the capping ability that the polyphenol extract matrix induces on the nZVIs, prolonging their reactivity (Nadagouda *et al.*, 2010); (iii) the valorization of natural products that, in some cases, are considered wastes or do not have any added value; and (iv) the fact that the extracts have high water solubilities, low toxicities and can act as a nutrient source which can enhance complementary biodegradation (Hoag *et al.*, 2009). Nadagouda *et al.* (2010) compared the nZVIs produced using the green and borohydride methods and observed that some of the green nZVIs were found to be nontoxic when compared with control samples prepared using the conventional borohydride reduction protocols.

References

Li, X.Q., Elliott D.W. and Zhang, W.X. (2006) Zero-valent iron nanoparticles for abatement of environmental pollutants: materials and engineering aspects. *Critical Reviews on Solid State Material Sciences*. **31**: 111-22

Kuhn, L.T., Bojesen. A. Timmermann, L., Nielsen, M.M., Morup, S. (2002) Structural and magnetic properties of core-shell iron-iron oxide nanoparticles. *Journal of Physics Condensed Matter* **14**: 13551-6.

Karlsson, MNA., Deppert, K., Wacaser, B.A., Karlsson, L.S., Malm, J.O. (2005) Size-controlled nanoparticles by thermal cracking of iron pentacarbonyl. *Applied Physics A: Material Science and Processing*; **80**:1579-83.

Wang, C.B., and Zhang, W.X. (1997) Synthesizing nanoscale iron particles for rapid and complete dechlorination of TCE and PCBs. *Environmental Science and Technology*; **31**: 2154-6

Hoag, G.E., Collins, J.B., Holcomb, J.L., Hoag, J.R., Nadagouda, M.N., Varma, R.S. (2009) Degradation of bromothymol blue by'greener' nano-scale zero-valent iron synthesized using tea polyphenols. *Jounral of Material Chemistry*; **19**: 8671.

Nadagouda, M.N., Castle, A.B., Murdock, R.C., Hussain, S.M., Varma R.S. (2010) In vitro biocompatibility of nanoscalezerovalent iron particles (NZVI) synthesized using tea polyphenols. *Green Chemistry*; **12**: 114-22.

Chapter 16

Nanoscale Characterization Techniques

Characterization Tools: Microscopy

Techniques	What it can do
Optical microscopy (OM)	Imaging and morphology of micron sized grain structure
Scanning near-field optical microscopy (SNOP)	investigation of nanostructures (uses property of evanescent waves)
Scanning electron microscopy (SEM)	Surface morphology /topography/imaging
Transmission electron microscopy (TEM)	Particle shape analysis and imaging (electron transmission through ultra-thin films and nanoparticles)
High-resolution transmission electron microscopy (HRTEM)	Imaging at atomic scale (crystal structures/defects)
Scanning transmission electron microscopy (STEM)	Electron beam focused into a narrow spot which is scanned over the sample- allows direct correction of image with quantitative analysis
Field emission microscopy (FEM)	High resolution surface morphology and surface crystallography-based on the work function difference of the various crystallographic planes of the surface
Selected area electron diffraction (SAED)	Information on local structure
Scanning tunneling microscopy (STM)	Surface topography and density of states information (scanning tunneling spectroscopy of the sample) (conductor/semiconductor) probed-uses the phenomenon of quantum mechanical tunneling between the nano-scaled tip and the surface

Techniques	What it can do
Atomic force microscopy (AFM)	Surface morphology/topography/Thickness measurement
Magnetic force microscopy (MFM)	Magnetic domain imaging

Characterization Tools: Spectroscopy and Other Techniques

Techniques	What it can do
Nuclear magnetic resonance (NMR) Spectroscopy	Detection of presence of particular nuclei in a material –the strength of the NMR signal is proportional to the number of resonating nuclei
Electron spin resonance spectroscopy (ESR)	Analysis of materials with unpaired spins- the basic concept of ESR are analogous to those of NMR. Energy change arises from the reversal of spin of the nucleus/electron (0.001^{-10} J ($mole^{-1}$)
Microwave spectroscopy	Study of rotating molecules (rotational spectroscopy). Separation between rotational levels of molecules are hundreds of joules ($mole^{-1}$)
Infrared spectroscopy	Vibrational spectroscopy –separation between levels are 10^4 J $(mole)^{-1}$
Fourier transform infrared spectroscopy (FTIR)	Asymmetrical vibration analysis. Fourier (mathematical) transform used to get actual spectrum from raw data.
Diffuse reflectance spectroscopy (DRS)	Reflectance and band-gap study. More intensity can be collected from the bulk of the sample in DRS study which is the main difference from the specular reflectance study
X-Ray Photoemission spectroscopy (XPS)	Surface chemical analysis; elemental composition, valence state of elements in a material. X-Rays (1-2 keV) are used and ultra-high vacuum Is required for XPS study
Raman spectroscopy	Used to observe vibrational, rotational, and other low frequency modes in a system. It relies on inelastic or Raman scattering of monochromatic light from a laser in the visible, near IR or near UV range
Energy Dispersive X-Ray (EDX) spectroscopy Wavelength dispersive X-Ray spectroscopy (WDS)	For elemental analysis (attached to electron microscope)
X-ray diffraction (XRD)	Crystal structure analysis

Small angle X-ray scattering (SAXS)	Using elastic scattering of X-rays (λ = 0.1 -0.2 nm) it can be used for study of density of materials, colloidal structures, particle sizes, porosity, domain sizes orientation, phase identification
X-ray Fluorescence (XRF)	Elemental analysis using secondary or fluorescent X-rays.
Electron energy loss spectroscopy (EELS)	Determination of elemental components and valence state of a materials by inelastic interaction (inner-shell ionizations)

XANES and XAFS

X-ray near-edge structure (XANES) also known as near-edge X-ray absorption fine structure (NEXAFS) is an absorption spectroscopy technique. The minor difference between XANES and NEXAFS lies in the energy range above the absorption edge. The X-Ray absorption fine structure (XAFS) deals with the absorption of X-rays by an atom at energies near and above the core level binding energies of an atom (Rao and Singh, 2013).

In photoelectric effect X-Ray is absorbed and a core-level electron is promoted out of the atom. X-ray fluorescence and the Auger effect give the probability of emission that is proportional to the absorption probability. The physical and chemical states of the atom modulate the atoms X-Ray absorption probability. XAFS spectra provide information about the formal oxidation state, coordination chemistry, distances and species of the atom immediately surroundings the selected elements. It is an interference effect between outgoing and backscattered waves requiring a coherent final state. It does not require long-range order but it does require short range correlations between atoms (liquids, amorphous solids). XAFS probes the path as far as the (elastically scattered) photoelectron can reach while the core hole is alive. The mean free path (λ) is determined by the core-hole lifetime and the inelastic losses. XAFS is an excited final-state effect. This means that the length and time scales are related.

X-Ray Photoelectron Spectroscopy

X-Ray Photoelectron spectroscopy (XPS) also known as electron spectroscopy for chemical analysis (ESCA) is a semi quantitative technique for determining composition based on the photoelectric effect. Detection of photoemission from materials helps us evaluate important electrical and optical properties. We know in a photoelectric process

Kinetic energy = hv-Ionozation potential (for gases)

Kinetic energy = hv-binding energy- work function (φ)
(for solids)

where, h = Planck's constant
v = Frequency

There is no photoemission if hv < φ and the kinetic energy of photoelectron increases as the binding energy (BE) decreases. The intensity of photoemission is proportional to the intensity of photons. The binding energy of the orbital is not affected by isotopes. XPS spectra show the characteristics "stepped" background. Due to inelastic process (extrinsic losses) from deep inside the bulk materials only electron close to surface can on an average escape without energy losses. Electron situated deep inside the surface lose energy and emerged with reduced kinetic energy and increased BE, whereas electron very deep inside the surface lose all the energy and can not escape.

Reference

Rao, M.S.R and Singh, S. (2013) Nanoscience and Nanotechnology: Fundamentals to Frontiers, Wiley India Pvt. Ltd, New Delhi 110002, pp. 361-363.

Chapter 17

Structure Determination by X-Ray Diffraction

X-ray diffraction (XRD) is an important experimental tool used in determining crustal structure of material system in bulk and nanorcgimes. Therefore, it is important to understand the basic concepts of XRD. Typical interatomic distance in a solid are of the order of few angstroms (1 Å=10^{-8} cm). An electromagnetic (EM) probe of the microscopic structure of a solid, therefore, should have a matching wavelength corresponding to the energy of the order of

$$E = h\nu = \frac{hc}{\lambda} = \frac{hc}{10^{-8}\,\text{cm}} = 12.3 \times 10^3 \text{ eV}$$

Energies of a few kilovolts or keV produce characteristic X-rays of wavelength ~1-2 Å. Wavelength of X-rays ($\lambda_{X\text{-ray}}$ ~1-2 Å) can be used to realize information about the crystal structure whose dimensions are in the same range as in the wavelength used. W.H. Bragg and W.L. Bragg found that in crystalline materials, characteristic pattern of diffracted X-Rays for certain well-defined wavelength and incident direction could be observed in forms of intense peaks of diffracted radiations. Bragg explained this on the basis of considering a crystal consisting of parallel planes of atoms that are held together by electrostatic forces in a crystalline solid.

Production of X-rays

Electron produced by thermionic emission, are made to accelerate between a high potential difference (20–40 kV), impringe on a metal target, a small percentage of their kinetic energy is converted into X-rays. The X-rays emitted by sources made of certain heavy elements (Mo, Cu, Co, Fe, Cr and Zr) consists of a continuous wavelength range which is termed as "white radiation" (Brensstrahlung) (Fig. 1).

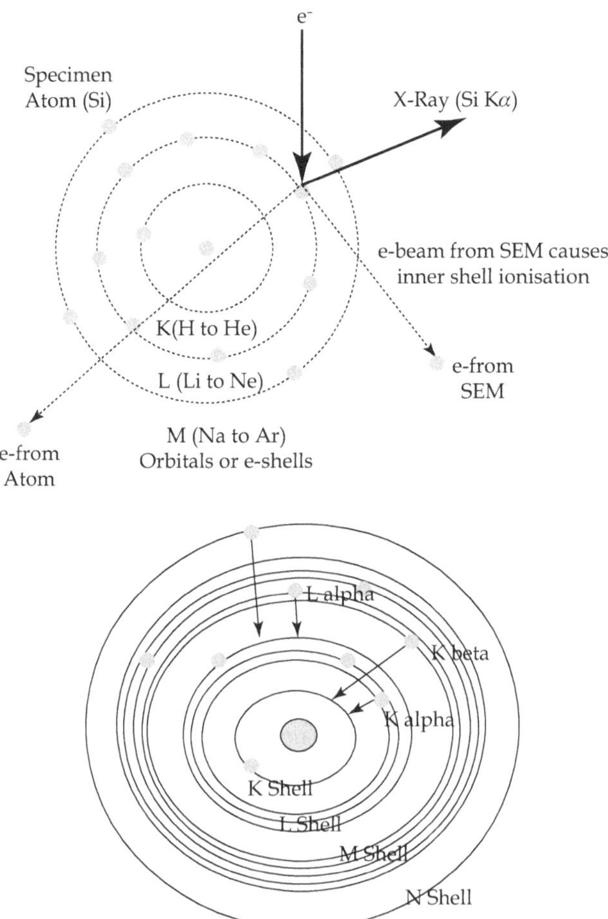

Fig. 1 Process of emission of X-Ray from Si atoms and basics of α, β radiations

The minimum wavelength in the continuous wavelength (λ) spectrum is inversely proportional to the applied voltage that accelerates the electrons toward the target. If the applied voltage is sufficiently high, then in addition to the radiation continuum, a characteristic radiation of a specific wavelength and intensity is also emitted by the target (Tarafdar and Raliya, 2012).

Target metals	Mo	Cu	Co	Fe	Cr
K± wavelength (Å)	0.71	1.54	1.79	1.94	2.29

Bragg's law

Bragg accounted for XRD by considering d, the interatomic separation between the lattice planes consisting of an array of atoms. Conditions for a sharp intense peak of the diffracted X-Ray beam are that: X-rays should be secularly diffracted by the atoms/ions in any one plane and there should be constructive interference between the diffracted rays originating from successive planes.

The path difference between the two rays ~ $2d \sin\theta$, where θ is the angle of incidence . For the waves to interfere constructively, this path difference must be an integral member of wavelengths and leading to the well known Braggs 'law

$$n\lambda = 2\, d \sin\theta$$

where n is an integer corresponding to the order of the reflection, it may be noted that $n = 1$ (first order) and $n = 2$ (second order).

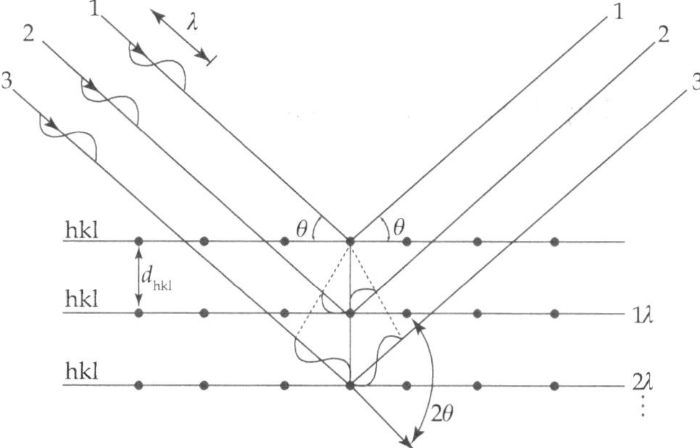

Fig. 2 XRD and Braggs's law

Scherrer Method

One of the major contributions in line width broadening comes due to reduction in particle size. There is certain range of particles sizes from 10^{-3} to 10^{-5} cm, where X-ray diffraction is quite insensitive to variations in grain size. However, as the grain size becomes less than 10^{-6} cm (100Å), we can call it precisely *"Particle size"*. Crystals in this size range can cause broadening of the debye rings that are produced by the diffraction of X-rays from the crystals of the specimen as determined by Braggs law. The empirical formula for obtaining the crystal size is given by

$$D = \frac{0.9\lambda}{B\cos\theta}$$

Where λ is the wavelength, B gloves the broadening of diffraction line measured at half of its maximum intensity in radians [also called full width at half maxima (FWHM)] (Fig. 3) and D is the average crystalline size.

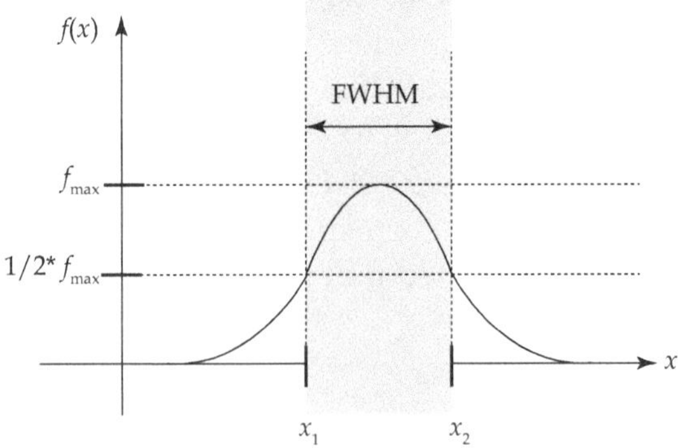

Fig. 3 Full width at half maxima (FWHM)

Particle Size Analyzer: Dynamic Light Scattering

Dynamic light scattering (DLS) also known as photo correlation spectroscopy or quasi elastic light scattering is a technique in physics, which can be used to determine the size distribution profile of small particles in suspensions or polymers in solution. It can also be used to probe the behaviour of complex fluids such as concentrated polymer solutions.

When light hits small particles the light scatters in all directions (Rayleigh scattering) so long as the particles are smaller than wavelength (below 250 nm). If light source is laser, and thus is monochromatic and coherent, then one observes time-dependent fluctuations in the scattering intensity. These fluctuations are due to the fact that the small molecules in the solutions are undergoing Brownian motions and so the distance between the scatters in the solution is constantly changing with time. This scattered light then undergoes either constructive or destructive interference by the surrounding particles and within this intensity fluctuations, information is contained about the time scale of movement of scatters. The DLS system measures particle size from 1 nm to 6 μ at concentrations 40% w/v. All DLS systems measure light scattering effects arising from the Brownian motion of particles in suspension (Tarafdar and Raliya, 2012).

Particle size distribution analyzers based on measuring the phenomenon of Brownian motion can be broadly classified as being based on either autocorrelations or on power spectrums. For the purpose of this documents, systems using autocorrelations will be called Photon Correlation Spectroscopy (PCS) systems. Nanoparticle size analyzers uses the DLS method to measure the size distribution of particles undergoing Brownian motion.

DLS can measure suspensions and emulsions from 1 nm to 1 μm. Both upper limit and lower limits are sample dependent. The lower limit is influenced by concentration and how strongly particles scatter the light. A low concentration of weakly scattering particles near 1 nm can be extremely difficult to reproduce. The upper size limit is determined mainly by density of the particles. DLS algorithms are based on all particle movements coming from Brownian motion. Motion due to scattering is not interpreted correctly by DLS systems. In additions, particles settled on the bottom of the sample cuvette can not be inspected by the laser light source. Particles with higher density will settle more quickly than low density particles. The upper limit of DLS may be 6 μm for emulsion samples where the two phases have similar density.

Serious DLS work could involve a dilution study to determine the nature of the particle-particle interactions and presence of multiple light scatterings. More sophisticated DLS system can also measure the sample characteristics including zeta potential, molecular weight and second virial coefficients.

The dynamic information of the particles is derived from an autocorrelation of the intensity trace recorded during the experiment. The second order auto-correlation curve is generated from the intensity trace as follows:

$$g^2(q;r) = \frac{(I(t)I(t+\tau)}{(I(t))^2}$$

Where, $g^2(q; r)$ is the auto-correlation function at a particular wave vector, q, and delay time, τ and I is the intensity. At short time delays, the correlation is high because the particles do not have a chance to move to a great extent from initial state that they were in. The two signals are thus essentially unchanged when compared after only a very short time interval. As time delays become longer, the correlation decays exponentially, meaning that after a long time period has elapsed, there is no correlation between the scattered intensity of the initial and final states. This exponential decay is related to the motion of the particles, especially to the diffusion coefficients. To fit the decay (*i.e.*, autocorrelation function), numerical methods are used, based on calculations of assumed distributions.

If the sample is monodisperse then the decay is simply a single exponential. The Siegert equation relates the second order auto-correlation function with the first order auto-correlation function g (q;r) as follows

$$g^2(q; r) = 1 + \beta\, [g^1(q; r)]^2$$

where, the parameter β is a correlation factor and depends on the geometry and alignment of the laser beam in the light scattering set up. It is roughly equal to the inverse of the number of speckles from which light is collected. The most important use of the auto = correlation function is its use for size determinations (Tarafdar and Raliya, 2012).

Basic equation followed

Stoke-Einstein equation

Translational diffusion coefficient determines the velocity of particles under Brownian motion. The translational diffusion coefficient can be converted into particle size using the Einstein-Stokes equation.

$$Dh = \frac{KbT}{3\pi\eta Dt}$$

Dh = Hydrodynamic radius

Kb = Boltzmann's constant

T = Temperature

η = coefficient of viscosity

Dt = Diffusion coefficient

Instrumentation

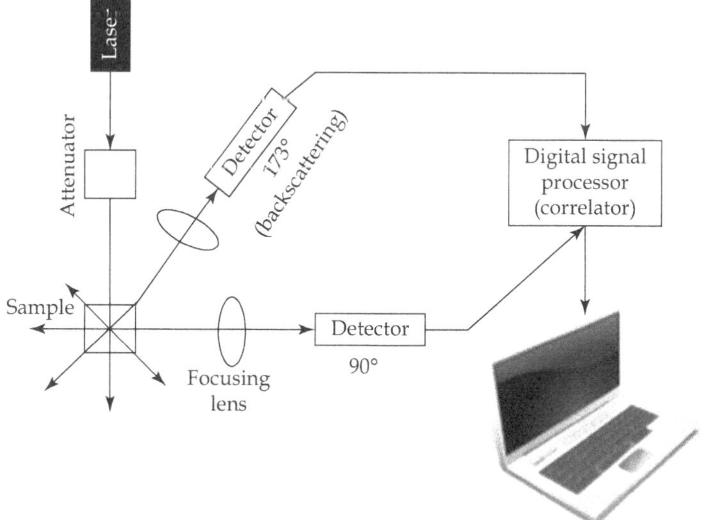

Fig. Instrumentation set-up for particle size analyzer

Reference

Tarafdar, J.C. and Raliya, R. (2012). The Nanotechnogy. Scientific Publisher (India). pp 87–100.

Chapter 18

Interaction of Electron with Matter

An electron impringing on a solid surface may be scattered once, several times, or never. When an incident electron beam bombard atoms of the sample, secondary electrons (SEs) and backscattered electrons (BSEs) emerge from the sample surface. Multiple scatterings occur if the dimension of the material is greater than approximately twice the mean free path of the electrons. This scattering may be elastic or inelastic. A pear shaped volume (interaction volume) is formed as shown in Fig. 1 and this volume of excitation increases with increase in the beam voltage (electron energy) and decreases with the increasing specimen density (ion atomic number).

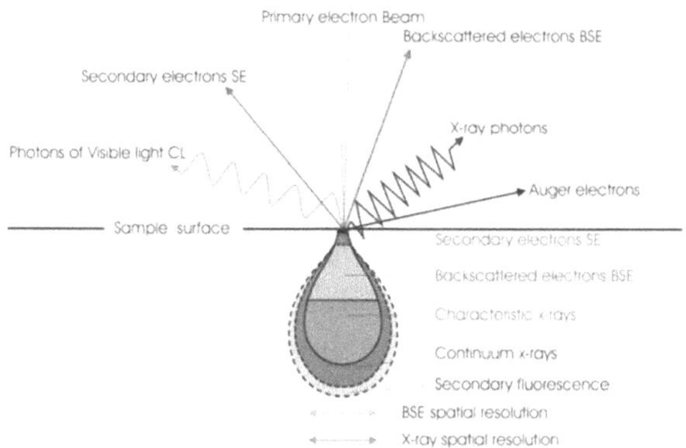

Fig. 1 Interaction of electron with matter

Scanning Electron Microscopy (SEM)

The SEM consists of a succession of lenses producing a finely focussed electron spot on the specimen. Various component of SEM are as follows (Fig. 2):

1. Electron optical column including the electron sources, magnetic lenses to demagnify the beam, magnetic coils to modify the beam and apertures to define the beam.
2. Vacuum system
3. Signal detection and display unit with detector and electronics.

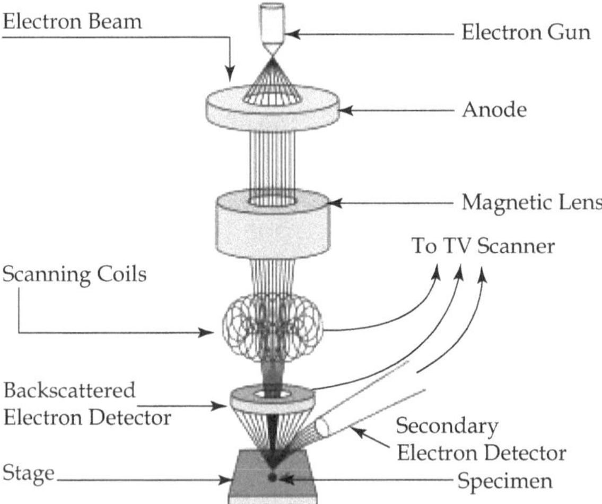

Fig.2 Schematic diagram of scanning electron microscope (SEM)

A SEM consists of a succession of magnetic lenses producing a finely focussed electron beam on the specimen. When an electron moves through this magnetic field it experiences the Lorentz force. This lens action is similar to that of an optical lens, in which a ray parallel to the axis of the lens is bent to the lens axis at the focal length of the lens. Focal length in an electromagnetic lens depend on following two factors:

1. The gun voltage (determining the electron velocity).
2. The amount of current through the coil (determining the flux density)

The SEM causes the electron beam to scan the sample surface. The emitted electrons are collected in form of current, amplified and made to fall on a cathode ray tube. The electron spot and cathode ray tube spot are scanned across the specimen and the tube surface respectively giving rise to an image of the sample surface topography.

Transmission Electron Microscope (TEM)

A transmission electron microscope (TEM) works like a slide projector with electron instead of light. Air is first pumped out to create vacuum. Then the electron gun emits a beam of high-energy electrons which is focussed *via* lenses. The beam strikes the sample and some electrons are transmitted. The transmitted electrons are focussed and amplified. The image contrast is enhanced by blocking out high-angle diffracted electrons and the image is passed though lenses for enlargement. When the image hits a phosphor screen, light is generated. The resolution of TEM is less than 1 nm. The essential components of TEM are presented in Fig. 3.

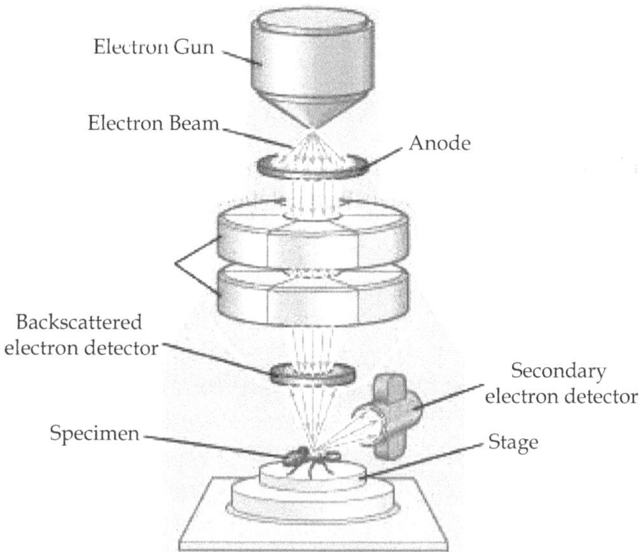

Fig. 3 Schematic diagram of transmission electron microscope (TEM)

Energy–Dispersive X-Ray (EDX) Analysis

Chemical composition of the sample can be analysed using energy dispersive X-ray (EDX) analysis facility attached to an SEM or TEM. When an incident electron beam bombards atoms of sample, SEs (secondary electrons) and BSEs (back scatteted electrons) are emitted from the sample surface. When the incident beams bounces through the sample creating SEs, it leaves thousands of atoms in the sample with holes in the electron shells where the SEs used to be. If these holes are in inner shells, electron from outer shells make transitions into the inner shells. However, the outer shells are at a higher energy states and therefore the atoms lose energy corresponding to the energy difference between two states (Fig. 4a). The energy difference usually lies in the region corresponding to X-ray.

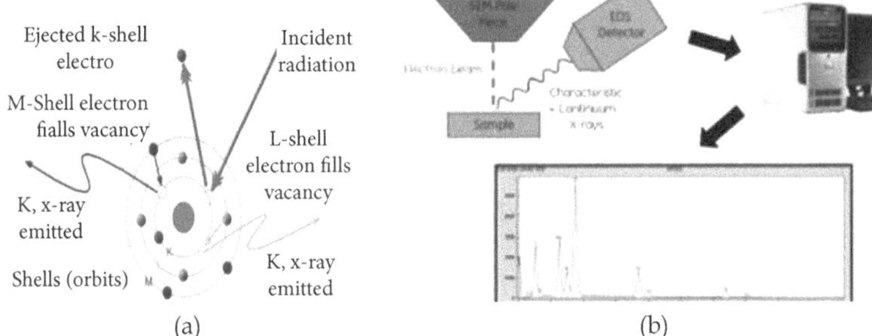

Fig. 4 Schematic diagram of EDX. Process of x-ray emission (a) and instrumentation (b)

The X-ray emitted from the sample atoms are characteristics, in energy and wavelength, not only of the elements of the parent atoms, but also of shells that have lost electrons. They are also the characteristics of those shells that have replaced them. Essentially, each elements has a characteristic X-ray line that allows the identification of the elemental composition of the sample and is the principle of EDX analysis. Area under the peaks of each identified elements are calculating after are taking into account the accelerating voltage of the beam to produce the spectrum. The area under the peak are then converted into weight or atomic percent that quantifies the composition of the constituting elements (Fig. 4b). However, the disadvantages of this method is that heavy elements with closer Z-values can not be easily distinguished.

Atomic Force Microscopy

Atomic force microscopy (AFM) is a nanoscale surface topography analysis technique. It can be effectively used to map the surfaces of insulating, semiconducting and metallic samples. A sharp tip (fabricated using lithography technique) of atomic dimension is located at the free end of cantilever (100-200 μm long). Typical atomic forces of covalently and ionically bonded atoms are $\sim 10^{-9}$ N and the AFM tips with sensitive spring constants offer forces $< 10^{-9}$ N on the surface, thus making imaging possible without causing any surface damage.

The measured cantilever deflections in AFM apparatus can be detected using a laser beam that falls on the backside of the relative-coated cantilever and the reflected laser beam intensity is measured using a photosensitive diode (split photodode) that in turn allows a computer to generate a map of surface corrugations (morphology/topology) (Fig. 5). AFM actually measures contours of constant attractive or repulsive forces. High aspect ratio AFM tips are micro fabricated from Si or Si_3N_4.

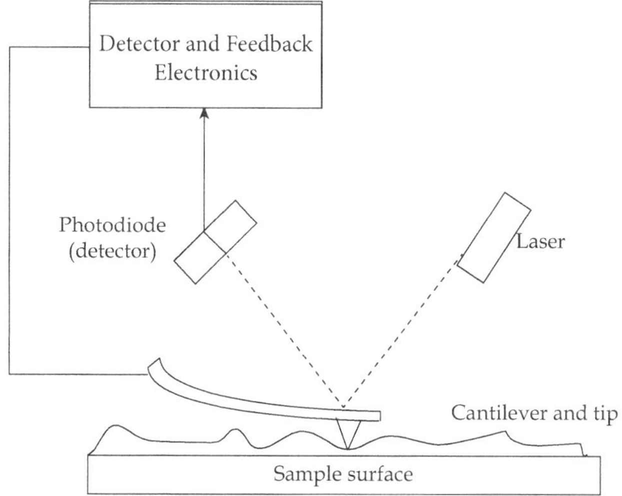

Fig. 5 Schematic Diagram of Atomic Force Microscopy (AFM)

The force is measured indirectly by measuring the deflection of the cantilever and knowing the stiffness of the cantilever. Hooke's law states that force $F = -kx$, where k is the stiffness of the cantilever and x is the extent to which the cantilever is bent.

Mode of Operation

Basically there are /modes of operation two contact and nanocontact mode. There is another category known as tapping mode. Out of these, contact mode is most widely used. The tip is scanned across the surface and it deflects as it moves over the surface corrugations. The tip is adjusted constantly to maintain a constant deflection and hence it maintains a constant height above the surface. This adjustment of the tip is displayed as surface information data. However, the ability to track the surface in this manner effectively is limited by the feedback circuit. As the tip is in hard contact with the surface it is required that the stiffness of the cantilever is less than that the effective spring constant holding the atoms together, which is of the order of 1-10 nN/nm. Most constant mode lever have a spring constant of ≤ 1 N/m.

Uses

Using AFM tips, it is possible to get force-distance curves. The slope of force-distance curve is measured which is then correlated with the elasticity of the sample. The data can be acquired along with topography which allows comparison of both height and material properties.

Infrared Spectroscopy

Infrared portion of electromagnetic radiation includes wavelength (λ) between 2.5 μm and 25 μm. Radiations in the vibrational infrared region of electromagnetic spectrum are expressed in terms of a unit called wavenumber (V–) rather than wavelength. Wavenumbers are expressed as reciprocal centimeters (cm^{-1}) and are easily computed by taking the reciprocal of wavelength expressed in cm.

$$V(\text{cm}^{-1}) = \frac{1}{\lambda(\text{cm})}$$

The main reason chemists prefer to use wave numbers as units is that they are directly proportional to the energy (a higher wave numbers correspond to higher energy). Thus in terms of wave numbers the vibrational infrared extends from about 4000 to 400 cm^{-1}.

Region of spectrum	Energy transitions
X-Rays	Bond breaking
Ultaviolet/visible	Electronic
Infrared	Vibrational
Microwave	Rotational
Radio frequencies	Nuclear spin (Nuclear magnetic resonance) Electron spin (Electron spin resonance)

Use of infrared spectra

Stretching vibration	Wave number (cm^{-1})
C-Cl C-O C-N C-C	650-1550
C=N C=C	1550-1650
C=O	1650-1800
C ≡ C C ≡ N	2000-2500
C-H N-H O-H	2500-4000

Modes of Stretching and Bending

The simplest types IR active of vibrational motion in a molecule, which gives rise to IR absorption, are the stretching and bending modes.

In general, asymmetric stretching, vibrations occur at higher frequencies than symmetric stretching vibrations, and all stretching vibrations occur at higher frequencies than bending vibrations.

Transmittance (T%)

Generally wavenumbers Vs light transmitted is plotted as IR spectra. This is recorded as percent transmittance (%T) because the detector records the ratio of the intensities of two beams

$$\%T = \frac{Is}{Ir} \times 100$$

Where Is and Ir are the intensities of the sample respectively. In many parts of the spectrum the transmittance is nearly 100% meaning that the sample is nearly transparent to radiation of that frequency (does not absorb it). Maximum absorption is thus represented by a minimum to the chart. Even so, the absorption is traditionally called a peak.

Fourier Transform Infrared Spectrometer (FTIR)

An interferogram is essentially a plot of intensity versus time (a time domain spectrum). However a chemist is more interested in a spectrum that is plot of intensity versus frequency (a frequency domain spectrum).

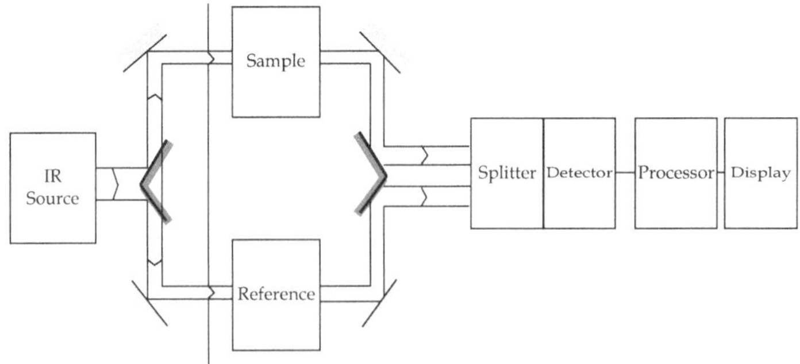

Fig. 6 Schematic diagram of FTIR

A mathematical operation known as Fourier Transform (FT) can separate the individual absorption frequencies from the interferogram producing a spectrum virtually identical to that obtained in dispersive spectrometer. This type of instrument is known as Fourier Transform Infrared Spectrometer (FTIR) (Fig. 6).

Microscopic techniques : Comparison

Microscope	Resolutions	Depth of Field	Interaction/ Matter	Environment
Optical (Photons)	~0.3- 1μm	Low 1 μm@ 10^3 ×	Light (electro-magnetic wave), insulating vs. conductor	Transparent, air, gas, vacuum, liquids
Scanning Electron Microspore (SEM)	1-10nm	Large ~mm @ 10^3 ×	Charged particle/ electric (+ magnetic) field, secondary+ backscattered e-, interband transition + ionization	Vacuum (+ recently gas p. 50 mbar)
Transmission Electron Microscope (TEM)	0.2 nm	Large but thin sample	Charged particle/ electric (+ magnetic) filed, e scattering and diffraction	Vacuum (gas or liquid in environmental cells)
Scanning Tunnelling Microscopy (STM)	Lateral 0.2 nm Depth 0.01 nm	Very low 0.01 nm	e- tunnelling, density of empty electronic state	Dielectric media; vacuum, air, Gas and liquid
Atomic Force Microscope (AFM)	Lateral ~ 1 nm Depth 0.1 nm	Very low 0.01 nm	Atomic forces (Van darwaals, covalent, ionic), friction, electrostatic+ magnetic forces	vacuum, air, Gas and liquid

Unit

Yotta (Y)	10^{24}	Deci (d)	10^{-1}
Zetaa (Z)	10^{21}	Centi (c)	10^{-2}
Exa (E)	10^{18}	Milli (m)	10^{-3}
Peta (P)	10^{15}	Micro (μ)	10^{-6}
Tera (T)	10^{12}	Nano (n)	10^{-9}
Giga (G)	10^{9}	Pico (p)	10^{-12}
Mega (M)	10^{6}	Femto (f)	10^{-15}
Kilo (k)	10^{3}	Atto (a)	10^{-18}

Hecto (h)	10^2	Zepto (z)	10^{-21}
Deca (da)	10^1	Yocto (y)	10^{-24}

Wavelength Scale in Energy in Electromagnetic Spectrum

Region	Wavelength (Å)	Wavelength (nm)	Energy (eV)
Radio	$>10^9$	$>10^8$	$< 10^{-5}$
Microwave	$10^6 - 10^9$	$10^5 - 10^8$	$10^{-5} - 0.01$
Infrared	$7000 - 10^6$	$700 - 10^5$	0.01 - 2
Visible	4000 - 7000	400-700	2 - 3
Ultaviolet	10 - 4000	1 - 400	$3 - 10^3$
X-Rays	0.1 - 10	0.01 - 1	$10^3 - 10^5$
Gamma rays	< 0.10	< 0.010	$> 10^5$

A simple formula to convert wavelength (micrometers) to energy in eV is $E(eV) = \dfrac{1.24}{\lambda(\mu m)}$

Physical Constants

Quantity	Symbol	Value
Angstrom unit	Å	1 Å$=10^{-4}$µm $= 10^{-8}$ cm
Avogadro constant	N	6.02204×10^{23} (mol)$^{-1}$
Boltzman constant	k_B	1.38066×10^{-23} J K^{-1}
Elementary charge	q	1.60218×10^{-19} C
Electron volt	eV	1 eV $= 1.60218 \times 10^{-19}$ J
Planck's constant	h	6.62617×10^{-34}Js
Speed of light in vacuum	c	2.99792×10^8 ms^{-1}